International Conference on Maritime Technology and its Applications (ICOMTA 2020)

AA007415

IOP Conference Series: Materials Science and Engineering
Volume 1175

Surabaya, Indonesia
12 December 2020

Editors:

Sryang Tera
S. Miftachudin

ISBN: 978-1-7138-4635-2
ISSN: 1757-8981

Printed from e-media with permission by:

Curran Associates, Inc.
57 Morehouse Lane
Red Hook, NY 12571

Some format issues inherent in the e-media version may also appear in this print version.

This work is licensed under a Creative Commons Attribution 3.0 International Licence.
Licence details: http://creativecommons.org/licenses/by/3.0/.

No changes have been made to the content of these proceedings. There may be changes to pagination and minor adjustments for aesthetics.

Printed with permission by Curran Associates, Inc. (2022)

For permission requests, please contact the Institute of Physics at the address below.

Institute of Physics
Dirac House, Temple Back
Bristol BS1 6BE UK

Phone: 44 1 17 929 7481
Fax: 44 1 17 920 0979

techtracking@iop.org

Additional copies of this publication are available from:

Curran Associates, Inc.
57 Morehouse Lane
Red Hook, NY 12571 USA
Phone: 845-758-0400
Fax: 845-758-2633
Email: curran@proceedings.com
Web: www.proceedings.com

TABLE OF CONTENTS

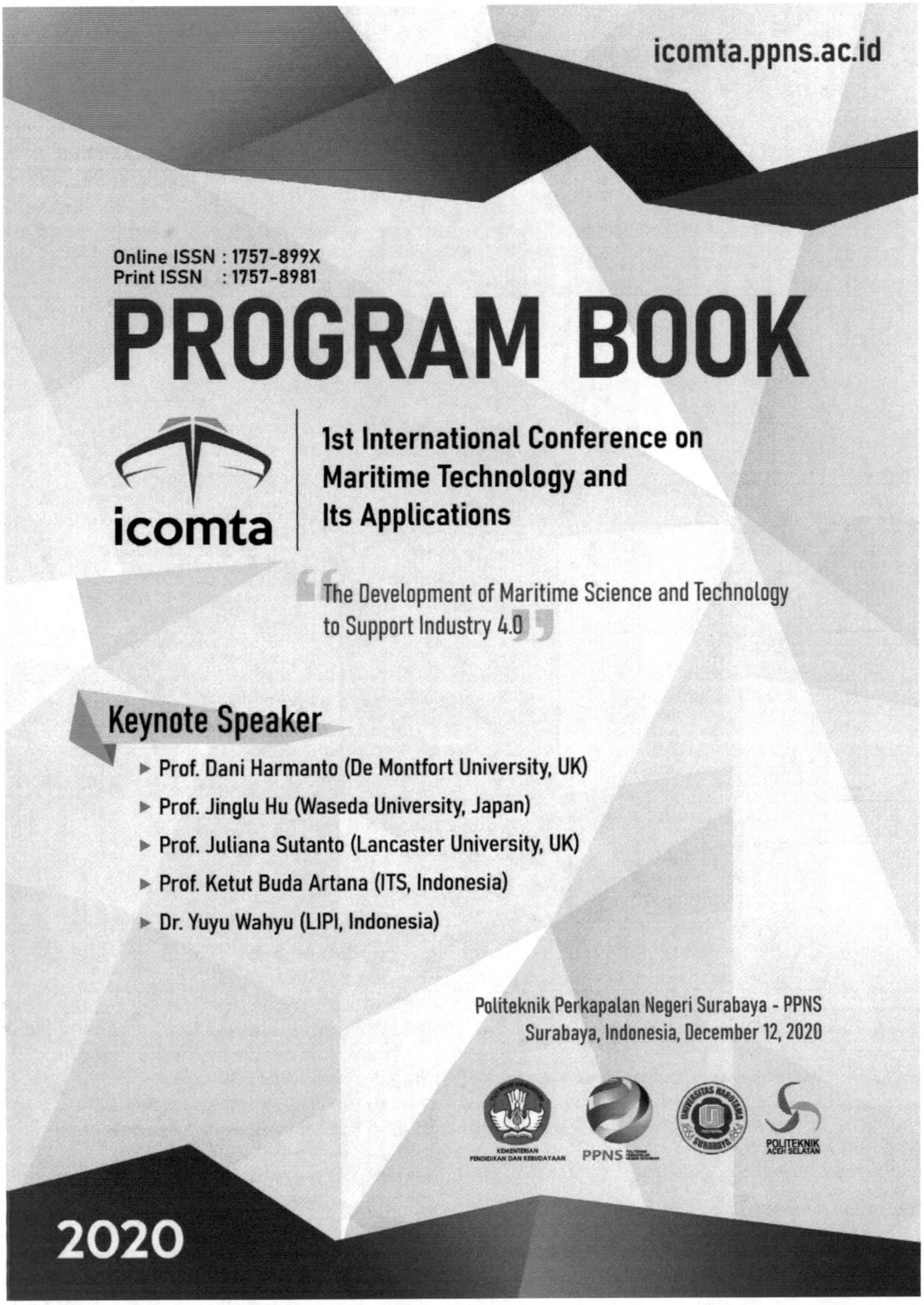

Content from this work may be used under the terms of the Creative Commons Attribution 3.0 licence. Any further distribution of this work must maintain attribution to the author(s) and the title of the work, journal citation and DOI.
Published under licence by IOP Publishing Ltd

IOP Conf. Series: Materials Science and Engineering 1175 (2021) 011001 doi:10.1088/1757-899X/1175/1/011001

icomta

1st International Conference on Maritime Technology and Its Application (ICOMTA) 2020

Organizing Committee

Advisory	Ir. Eko Julianto, M.Sc., FRINA
	Dr. Eng. Muhammad Anis Mustaghfirin
Chairman	Dr. Eng. Mohammad Abu Jami'in
Secretary	Ryan Yudha Adhitya, S.ST, MT
Finance	Novi Eka Mayangsari, ST., MT.
Editorial	Sryang Tera Sarena, ST., M.Sc.
	Miftachudin, S.Pd., MPd.
Publication	Yugowati Praharsi, Ph.D
	Dr. Eng. Imam Sutrisno
IT team	M. Khoirul H., S.Kom., M.Kom
	Afif Zuhri A., ST., MT.
Industrial Sponsors	Dika Rahayu Widiana, PhD
	Haidar Natsir A., S.ST, MT.
Equipment Team	Mades Darul Khairansyah, S.ST, MT.
	Priyambodo Nur A., PhD.

Committee

Prof. Dani Harmanto	De Montfort University, United Kingdom
Prof. Benny Tjahjono	Coventry University, United Kingdom
Dr. Taufiq Asyhari	Cranfield University, United Kingdom
Dr. Irwanda Laory	Warwick University, United Kingdom
Prof. Ei-Ichi Kobayashi	Kobe University, Japan
Prof. Jinglu Hu	Waseda University, Japan
Prof. Vikas Kumar	Bristol University, United Kingdom
Prof. Hui-Ming Wee	Chung Yuan Christian University, Taiwan
Prof. Ming-Jyh Chern	NTUST, Taiwan
Dr. Muhammad Ikhsan S.	Universitas Narotama Surabaya, Indonesia
Ali Ridlo Barakbah, PhD	PENS, Indonesia
Dr. Viv Djanat Prasita	Universitas Hang Tuah, Indonesia
Dr. Wahyu Mulyo Utomo	Universiti Yun Hussein Onn Malaysia, Malaysia
Prof. Taufik	Cal Poly State University, California, USA
Prof. Makbul A.M. Ramli	King Abdulaziz University, Saudi Arabia
Prof. Wisnu Jatmiko	Universitas Indonesia, Indonesia
Prof. Hamiruce Marhaban	Universiti Putra Malaysia, Malaysia
Prof. Joni Hermana	Institut Teknologi Sepuluh Nopember, Indonesia
Prof. Ketut Buda Artana	Institut Teknologi Sepuluh Nopember, Indonesia

IOP Conf. Series: Materials Science and Engineering 1175 (2021) 011001 doi:10.1088/1757-899X/1175/1/011001

icomta

Welcome Message from Chairman of ICOMTA 2020

Welcome, Ladies and Gentlemen,

As chairman of ICOMTA, I have the honor to welcome all the keynote speakers, presenters, and participants to the first International conference on maritime technology and its applications (ICOMTA 2020). I wish to express our satisfaction to the organizers of the conference, through cooperation and hard work, we can come together to this event.

The theme for this conference is "The Development of maritime science and technology to support industry 4.0".

This theme was taken by considering that PPNS is one of the vocational educations that is consistent in developing technology in maritime field. PPNS is an institution of higher vocational-education that provides a dynamic and progressive learning environment producing skilled and professional graduates in their knowledge background. Therefore, supporting green development focused on maritime sciences and technology through international collaboration, networking, sharing experience and knowledge is very important to improve academic and research with global or international recognition. It is also as a vision of PPNS that becomes a polytechnic with global reputations. We wish that ICOMTA conference is the one international activity with many impacts to contribute the development of maritime science and technology for PPNS and all over the worlds.
I sincerely hope that you will enjoy the conference and thank you for your participation.

Warm regards,

Dr. Eng. Mohammad Abu Jami'in
Chairman

IOP Conf. Series: Materials Science and Engineering 1175 (2021) 011001 doi:10.1088/1757-899X/1175/1/011001

icomta

Welcome Message from Director of PPNS

Welcome, Ladies and Gentlemen,
Dear Colleagues,
I am pleased to welcome you to the first International Conference on Maritime Technology and its Application 2020. It is high time that such an event took place to highlight the maritime technology and its applications. It is a good topic for researchers, scientists and the entire engineering community to meet each other and sharing their thoughts, knowledge and ideas. We also intend to bring together industrial practitioners in the maritime field and decision makers to share experiences and discuss maritime technology and its development. But this year, this conference holds special value because this is the first international conference organized by PPNS where the entire conference is taking place on a digital platform. Due to the changing time caused by the COVID-19, we encourage our participants to stay safe and maintain social distancing as a result of which this conference will be the first of many conferences in the future that will be conducted remotely through the use of advanced technologies such as augmented reality and virtual reality.
We carry out this international conference with good cooperation between the organizing committee, committee, keynote speakers, writers and participants, reviewers and the technical Committee. Let me take this opportunity to express my sincere appreciation to all of you. Even in the atmosphere of COVID-19, I would like to encourage the delegation to participate in all interesting activities and discussions via online platforms during the conference. We can share experiences and useful knowledge through this conference. I am sure you will have fruitful and rewarding exchanges in the next few days. I wish you every success with this important conference and I look forward to learning about the outcome.

Thank you.

Ir. Eko Julianto, M.Sc., FRINA
The Director of PPNS

IOP Conf. Series: Materials Science and Engineering 1175 (2021) 011001 doi:10.1088/1757-899X/1175/1/011001

icomta

Keynote Speakers

Prof. Jinglu HU received a Ph.D degree in Computer Science and System Engineering from Kyushu Institute of Technology, Iizuka, Japan in 1997. From 1986 to 1988, he worked as a Research Associate and from 1988 to 1993 a Lecturer at Sun Yat-Sen University. From 1997 to 2003, he worked as a Research Associate at Kyushu University. From 2003 to 2008, he worked as an Associate Professor and since April 2008, he has been a Professor at Graduate School of Information, Production and Systems of Waseda University. His research interests include Computational Intelligence and its applications to system modeling and identification, bioinformatics, time series prediction, financial analysis, data mining and so on.

Ketut Buda Artana is a professor at Marine Engineering Department of Sepuluh Nopember Institute of Technology (ITS). He completed his bachelor degree in the Marine Engineering program at ITS in 1993; Master of Marine Engineering at Newcastle University in 1997; and Ph.D in the Marine Engineering Program at Kobe University in Japan, 2003.He has dedicated his academic career life for teaching and professional development starting from becoming Vice Rector for Innovation, Partnership, Alumni and Internationalization. His next remarkable dedication is manifested as Head for Center of Excellence for Maritime Safety and Marine Installation and also Member of Technical Advisory Board of Indonesian Classification Society. In 2018, he was granted an award on Innovation as Best Start-up at Ministry of Technological Research and Higher Education and achieved Academic Leader Award in Maritime Field in the following year. His contribution to the field of Marine Engineering come up through visiting professor or guest lecturer program at three outstanding universities in Japan (i.e. Kyushu, Kobe and Kumamoto University), NTUST Taiwan and UTM Malaysia. His incredible insight and expertise are reflected through several books publication and achievements on Intellectual Property Right.

icomta

Dr. Yuyu Wahyu

LIPI, Indonesia

Yuyu Wahyu (Member, IEEE) was born in Bandung, Indonesia, in February 1962. He received the Ir. (Insinyur) degree, the M.Eng. (M.T.) degree in telecommunication information system from electrical engineering study program, and the Ph.D. degree in global information and telecommunication studies from the School of Electrical and Informatics Engineering, Institut Teknologi Bandung, Bandung, Indonesia, in 1990, 2000, and 2010, respectively. He has been with the Telecommunications Research Center, Strategic Electronics, Components and Materials (Telkoma) Indonesian Institute of Sciences (LIPI) (now the Research Center for Electronics and Telecommunications, LIPI), since 1991. He was the Head of the Telecommunications and Radio Laboratory, from 2000 to 2003, and research facilities, from 2010 to 2016. He has been the Chair of the Research Group of Antennas and Propagation, since 2014. Since 2019, he has been appointed as a Research Professor in telecommunications transmission. He followed a number of activities related to his field of competence, including Guest Research in Okayama, Japan, in 2003, for 1.5 months on the topic of Active Antenna and Radar Training, in 2006 and 2007, at IRCTR-TU, Delft, The Netherlands. He conducted FMCW radar research from 2006 to present and electronic support measure (ESM), from 2015 to 2018. He participated in professional organizations, including: Himpenindo (Association of Indonesian Researchers), since 2018, the Indonesian Radar Association, since 2008, and the IEEE Antenna and Propagation Society, since 2010.

icomta

Dani Harmanto is an Associate Professor in Aeronautical Engineering, De Montfort University. His ultimate research interest and areas of expertise cover the following: Vehicle Aerodynamics, Healthy Building, Hydrodynamics (War Ship Design), Finite Element Analysis, Design Optimization, Product Development, Natural Material Development, Wind Turbine Application, Renewable Energy, and Lean Management. He graduated from National Institute of Technology, Malang, Indonesia and got the title of Ir. In 1993. He received his Msc. in Automotive Engineering Design and Manufacture from Coventry University, Coventry, UK in 1995, and his Ph.D from the same university in 1999. Taking part in the Advisory Boards of some reputable International Journals, he has reviewed many proceedings and journals of International conferences since 2009 under various theme: Engineering Technology, Science, Manufacturing, Energy, materials, robotics and mechatronics. He also reviewed the Book entitled “SolidWorks Simulation for Absolute Beginners by CRC Press/Taylor and Francis Group. He was awarded as Lecturer of the Year 2017 - 2018 by College of Engineering and Technology, University of Derby and was the Programme Leader for BEngs, Derby University (2007 - 2017). Recognized as the Honourable & Advisory Member of Lecturer Association of the Republic Indonesia since 2016, He was honoured to be the Managing Director of International Indonesian Scholars Association for UK (2018 - 2020).

IOP Conf. Series: Materials Science and Engineering 1175 (2021) 011001 doi:10.1088/1757-899X/1175/1/011001

icomta

Juliana Sutanto graduated from School of Computing, National University of Singapore (NUS) with a PhD degree in Information Systems. After graduation, she joined ETH Zurich in Switzerland as Chair of Management Information Systems. In August 2015, she moved to Lancaster University as Professor in Information Systems. Her research expertise is on system design, users' behavioral analysis (both quantitative and qualitative data), data management, and data visualization. She is a recipient of Informs ISS Design Science Award on privacy-safe design. She is currently developing research direction on societal innovation and digital governance in the context of smart parks, data-driven innovation, community resilience, and disaster management through active local and international stakeholders' engagement. She is a PhD. Director (previously was a Director of MSc. E-Business and Innovation) and Co-Director of the Connected Communities Research Lab (CCRL). In 2018, she co-authored report on Smart Parks (bringing smart technology to the green spaces) received wide media coverage such as the BBC and Sky News in UK, and abroad. In 2019, she co-authored another report on Greenside Lead Mine (Smart Heritage Destination) that also received wide media coverage. In early 2019, she co-organized a workshop in the United Nations Conference Centre in Bangkok within the framework of Accelerating Impact for Disaster Management and Emergency Response (AIDE-Response). The workshop was attended by representatives from ASEAN Coordinating Centre for Humanitarian Assistance (AHA Centre), disaster management authorities of the three most vulnerable countries in the region (Philippine, Indonesia, and Myanmar), NGOs, and UN agencies (FAO, UNESCAP, UNICEF). These reports and workshop are driving the development of my research on societal innovation and digital governance.

icomta

Conference Documentations

1. Conference Video could be accessed on the following Youtube page
   https://www.youtube.com/watch?v=SY4b-MYlJHQ

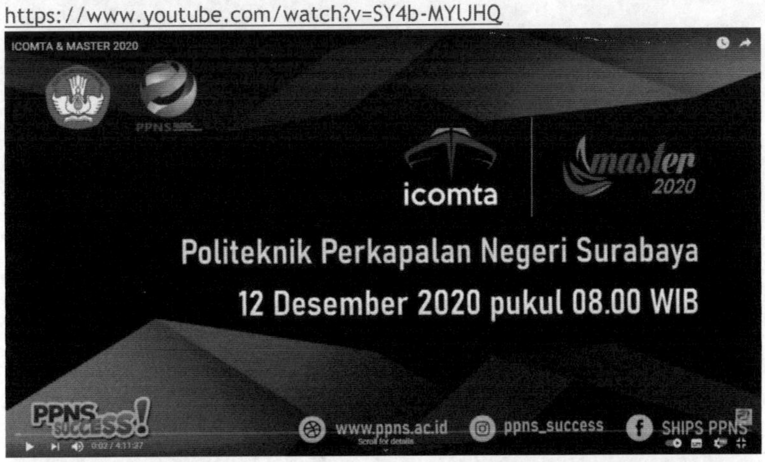

IOP Conf. Series: Materials Science and Engineering 1175 (2021) 011002 doi:10.1088/1757-899X/1175/1/011002

Peer review declaration

All papers published in this volume of *IOP Conference Series: Materials Science and Engineering* have been peer reviewed through processes administered by the Editors. Reviews were conducted by expert referees to the professional and scientific standards expected of a proceedings journal published by IOP Publishing.

- **Type of peer review:**
  Double-Blind Review

- **Conference submission management system:**
  www.icomta2020.edas.info

- **Number of submissions received: 28**

- **Number of submissions sent for review: 25**

- **Number of submissions accepted: 19**

- **Acceptance Rate (Number of Submissions Accepted / Number of Submissions Received X 100): 68%**

- **Average number of reviews per paper: 2**

- **Total number of reviewers involved: 30**

- **Any additional info on review process:**

  After the paper submitted through submission management system, the paper directly scrutinized by Editor to check for its scope, technical relevance, plagiarism, and originality using similarity software. Manuscripts which passed the criteria were sent to the field experts for Double-Blind Review.

- **Contact person for queries:**

  Dr. Eng. Mohammad Abu Jami'in
  Chairman
  Jammy@ppns.ac.id

Content from this work may be used under the terms of the Creative Commons Attribution 3.0 licence. Any further distribution of this work must maintain attribution to the author(s) and the title of the work, journal citation and DOI.
Published under licence by IOP Publishing Ltd

IOP Conf. Series: Materials Science and Engineering 1175 (2021) 012001 doi:10.1088/1757-899X/1175/1/012001

Emission characteristics of biodiesel fuels produced from Rapeseed Oil

M T Dieng[1], **T Iwanaga**[2], **Y C Yurie**[3] **and S Torii**[4]
[1,2,3,4]Department of Advanced Mechanical System Engineering, Kumamoto University, Japan

E-mail: torii@mech.kumamoto-u.ac.jp

Abstract. This present research is focused on emission characteristic of biodiesel fuels produced from rapeseed oil. Biodiesel fuels are produced through an optimization of transesterification conditions based on the variation of the molar ratio of alcohol /oil using methanol and ethanol. Best samples are obtained after a transesterification reaction involving an alcohol/oil molar ratio of 18:1, 1% of potassium hydroxide as catalyst, 60 minutes of reaction time, a reaction temperature at 60°C and a stirring speed of 650rpm. The biodiesel fuels show a higher fuel consumption rate and exhaust gas temperature when the engine speed is increased. They also exhibit lower NOx emission with a slight rise in CO and CO_2 emission compared to mineral diesel fuel tested in same conditions.

1. Introduction

Diesel engine constitutes the most effective internal combustion engine used in engineering machinery for power generation (Klaus Mollenhauer and Helmut Tschoeke, 2010). A vast majority of these engines is engaged in transportation (road, railway, naval etc.) and electricity generation, giving rise to their important role in the global economy moving (John, 2007). Fossil fuels are commonly used to run diesel engine due to its high thermal efficiency. The energy generated from the combustion of these fuels leads to an excellent engine performance. However, it simultaneously generates harmful emission gas such as CO_2 (carbon dioxide), CO (carbon oxide), NOx (nitrogen oxides), and PM (particular matters) which threat the sustainability of our ecosystem by contributing to the environmental pollution and global warming (Bentley, 2002). Additionally, the high energy demand combined with the limited petroleum is also major issues which lead to their fast depletion with expectation about 40 years (Gerling, 2007).

These are key factors that motivate research and development for alternative diesel fuel with maximum efficiency and minimum environmental pollutions. In this context, biodiesel can be considered as the most feasible options correspond to its renewability and compliance with fuel standard (Rafael and Juan, 2012).

Biodiesel can be obtained from biological sources such as vegetable oil, animal fat, algae and recycling cooking oil. These raw materials content triglycerides molecules which are converted into fatty acid alkyl ester (biodiesel) by using a specific reaction called transesterification. This is a chemical reaction involving triglycerides with an alcohol in the presence of a catalyst to yield biodiesel and glycerol as by-product (Mushtaq, 2013). The properties of the biodiesel depend strongly on this reaction and are strictly checked before its use in diesel engine. Nevertheless, the incomplete combustion of biodiesel in diesel engine releases harmful gas through the atmosphere. Among them,

Content from this work may be used under the terms of the Creative Commons Attribution 3.0 licence. Any further distribution of this work must maintain attribution to the author(s) and the title of the work, journal citation and DOI.
Published under licence by IOP Publishing Ltd

carbon monoxide (CO), carbon dioxide (CO_2), Oxides of Nitrogen (NOx) are widely emitted. In fact, many researches considered that CO emissions are reduced when diesel engine is fueled by pure biodiesel (Ozsezen et al., 2009). Experiments are carried out by Kalligeros et al. (2003) and Usta (2005) to estimate the emission characteristic of biodiesel fuels produced from sunflower oil and Tabaco seed oil, respectively. They report a lower CO emission of produced fuels. Alcohol type used in the biodiesel production can influence CO emission. In particular, Baiju et al. (2009) disclosed that biodiesel fuel based methyl esters emits less CO than those based on ethyl esters. In fact, biodiesel is a lower-carbon fuel, so it has a lower elemental carbon-to-hydrogen ratio than diesel (Ozsezen et al., 2009). On the contrary, Ramadhas et al. (2005) pointed out that CO_2 emissions from biodiesel are raised or are similar to diesel one, due to more efficient combustion from diesel fuels.

As for NOx emission, many researchers understand that the oxygen content in biodiesel is responsible of an increase tendency for NOx emission (for example, Hansen et al., 2006). However, Lapuerta et al. (2008) found that the oxygen content in biodiesel has no influence in NOx emission.
In order to produce higher quality biodiesel fuel for better estimation, the corresponding fuel properties can be controlled by the transesterification process. The purpose of the present study is performed to optimize the transesterification reaction conditions in order to obtain a quality biodiesel with good engine performance and low toxic gas emission. Best biodiesel samples from the optimum reaction conditions are tested in diesel engine in order to evaluate and compare their performance and emission characteristics with mineral diesel.

2. Experimental Method

2.1. Raw material and chemicals
Rapeseed oil namely Brassica Napus is selected as a raw material because it is widely available around the world and belongs to the preferred oil feedstock for large scale production of biodiesel (Hoekman et al., 2012). Methanol (CH_3OH; 99.8%) and Ethanol (C_2H_5OH; 99.5%) are used in transesterification reaction catalyzed by potassium Hydroxide KOH pellets (85.0%). Acetic Acid is used during the washing process performed after reaction.

2.2. Transesterification process
As a first step, 100g of rapeseed oil is poured in a glass flask of 500ml and preheated in thermal batch to eliminate moistures. Then, the required amount of alcohol (methanol or ethanol) and catalysts (potassium hydroxide), for each sample, is measured and mixed in an Erlenmeyer flask. The solution is stirred on magnetic plate to dissolve the KOH pellets to make methoxide/ethoxide solution which is then added into the preheated oil. The reaction is then started and is run for the required time under stirring condition at ambient pressure. Here, alcohol concentration is varied by making samples with 6:1; 9:1; 12:1; 15:1; 18:1 and 21:1 of molar ratio. Consequently, better transesterification parameters, for each experiment, are 1% of potassium hydroxide as catalyst; 60°C of reaction temperature; 30 min of reaction time and 650 rpm of stirring. After completion of reaction, samples are set to decantation for phase separation by gravity. In other words, the biodiesel floats on the top of the reaction vessel while the glycerol sinks in the bottom.

The crude biodiesel fuel produced is removed and treated with acetic acid to neutralize any unreacted hydroxide, methoxide or ethoxide. It is also washed with water to remove soluble salts (from potassium hydroxide), unreacted alcohols and partial glycerides. Any water that might have mixed with the biodiesel is removed by heating process.

2.3. Fuel Characterization
In general, the biodiesel fuel is characterized by its relative density, viscosity and high heating value.
The density is measured at 15°C according to EN ISO 12185 test method (Mert and Atilla, 2015).
Dynamic viscosity is measured by using a Brookfield digital viscometer (DV-II+ Pro) connected to a water bath with circulating pump (Brush less DC Pump, model DC 40-2470). Measurement is done at

IOP Conf. Series: Materials Science and Engineering 1175 (2021) 012001 doi:10.1088/1757-899X/1175/1/012001

40°C for three different share rate set at 7.4s-1, 15s-1 and 37s-1 based on the standard test method ATSM D445.

The calorific value is measured using the Shimadzu Auto-Calculating Bomb calorimeter CA-4AJ and according to ASTMD240-14 standard test method (Demirbas, 1998). It is calibrated for 26456 J/g at 20°C with standard benzoic acid. Here, measurement is repeated three times and its average is employed as the final value.

2.4. Engine Test

After analyzing the fuel properties of biodiesel produced here, best samples are tested in diesel engine to evaluate their performance and emission characteristic. Related engine characteristic of mineral diesel is measured under the same conditions to compare with that of produced biodiesel.

Engine test is performed by using Yanmar diesel engine (TF70V-E model). Its specification is given in Table.1.

Table 1. Specification of Diesel Engine Model TF70V-E

Measurement item	Measurement range	Resolution
CO_2	0…25Vol.%	0.01 Vol. %
CO, H_2-compensated	0…10000ppm	1ppm
NO	0…4000ppm	1ppm
NO_2	0…500ppm	0.1ppm
SO_2	0…5000ppm	1ppm
H_2S	0…300ppm	0.1ppm
CO_2-(IR)	0…50Vol.%	0.01 Vol. % (0~25Vol. %) 0.1 Vol. % (>25Vol. %)
HC	Natural gas: 100…40000ppm Propane: 100…21000ppm Butane: 100…18000ppm	10ppm 10ppm 10ppm
K-type (NCr-N)	-200~1370°C	0.1°C
S-type (Pt10Rh-Pt)	0~1760°C	1°C

Measurement is carried out at three different engine speeds, i.e., 1000, 1500 and 2000rpm. Each experiment is repeated three time to calculate the mean values and all data are obtained under steady operation. Fuel consumption rate is measured by observing the volumetric rate from the fuel tank supported by stopwatch.

The exhaust gas temperature and emission characteristic from the biodiesel are analyzed simultaneously by using a Testo 350 Flue Gas Analyzer (TESTO, Inc.). Flue gas analyzer port is set at the end of engine's chimney as show in the experimental set up (Figure 1). Note that the cetane number for the biodiesel fuel was not measured in the present study.

IOP Conf. Series: Materials Science and Engineering 1175 (2021) 012001 doi:10.1088/1757-899X/1175/1/012001

Figure 1. Location of Flue gas analyzer port.

3. Result and Discussion

In this section, the biodiesel fuels obtained via transesterification of rapeseed oil with methanol and ethanol are denoted as ROME (Rapeseed Oil Methyl Ester) and ROEE (Rapeseed Oil Ethyl Ester), respectively. In pre-experiment based on density, viscosity and heating value for different molar ratio and temperature, the optimal transesterification reaction conditions which give the best fuel properties are found to be an alcohol/oil molar ratio of 18:1, 1% of potassium hydroxide as catalyst, 30 minutes of reaction time, 60°C of reaction temperature and stirring speed of 650rpm. Two biodiesel types, i.e., ROME and ROEE produced under the above conditions, are summarized in Table. 2.

Table 2. Thermal properties for ROME and ROEE.

Samples	Relative density (15°C ; g/ml)	Kinematic viscosity (40°C; mm^2 /s; 15 $_s^{-1}$)	High heating value (MJ/kg)
ROME	0.865	3.49	40.81
ROEE	0.853	4.50	39.94
Diesel	0.806	2.50	45.44

Nitrogen oxides (NOx) are the most significant environmental pollutants emitted from a diesel engine. NOx emission under different engine speeds, form three fuels, is depicted in Figure 2. One observes that NOx decreases with the increase in engine speed. This is generally explained by the shorter residence time available for NOx formation, which may be the result of an increase both in the volumetric efficiency and flow velocity of the reactant mixture specifically at higher engine speed (Amit, 2012). At low and medium engine speed, diesel fuel produces maximum NOx compared to the two fuels, while at high engine speed (2000rpm), NOx emission from ROEE fuel

IOP Conf. Series: Materials Science and Engineering 1175 (2021) 012001 doi:10.1088/1757-899X/1175/1/012001

becomes slightly higher than the other tested fuels. The variation of NOx is proportional to the flue gas temperature and can also depend on the nature of the fuel. Generally, NOx emission includes nitric oxide (NO) and nitrogen dioxide (NO_2). NO is the predominant oxide of nitrogen generate from the engine combustion and contributes to ozone formation and is considered as non-toxic gas by itself.

..

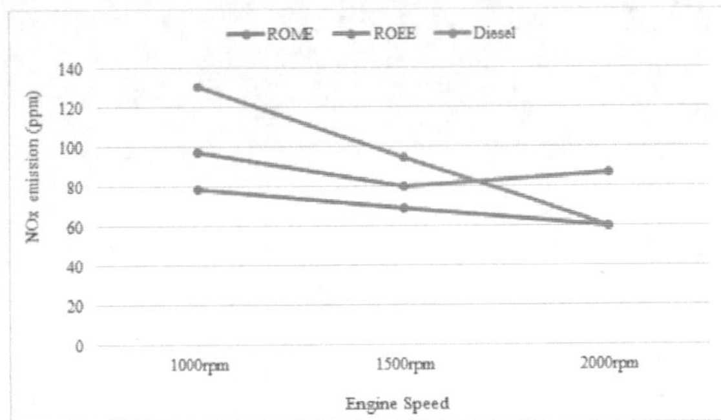

Figure 2. NOx emission from diesel engine under different speed.

Emission characteristic of CO, for biodiesel and diesel fuels, is depicted in Figure 3. CO emission from ROEE is continuously higher than other fuels and remains practically constant for any engine speed. This confirm that biodiesel based methyl esters emits less CO compared to those based on ethyl esters (Baiju et al., 2009). Meanwhile, the trends of CO emission from ROME and diesel are constantly un-stable with the variation of engine speed. In fact, when substituting diesel fuel with biodiesel many factors may influence significantly the variation of CO emission. Generally a decrease trend in CO emission is more evident even though some studies have found that CO emission from biodiesel is higher than of pure diesel fuel (Cardone et al., 2002; Krahl et al., 2009). At low engine speed, CO emission from biodiesel fuels is higher compared to diesel fuel. This is partly due to the poor atomization conditions influenced by the higher viscosity of biodiesel and the low flue gas temperature which lead to that high emission of CO (Fontaras et al., 2009). At medium engine speed, CO emission from ROME decrease and become the lowest compared to other fuels, as show in Figure 3.

Figure 4 illustrates CO_2 emission for biodiesel and diesel fuels. Emission from ROEE is lineally increased with an increase in engine speed. One observes that there is no significant variation of CO_2 emission from ROME under varied speed. In contrast, diesel fuel emits more CO_2 compared to ROME at medium speed before its emission decrease and reach the same level as ROME at high speed. The few carbon dioxide (CO) emission from ROME can be explained by its lower elemental carbon-to-hydrogen ratio than diesel.

IOP Conf. Series: Materials Science and Engineering 1175 (2021) 012001 doi:10.1088/1757-899X/1175/1/012001

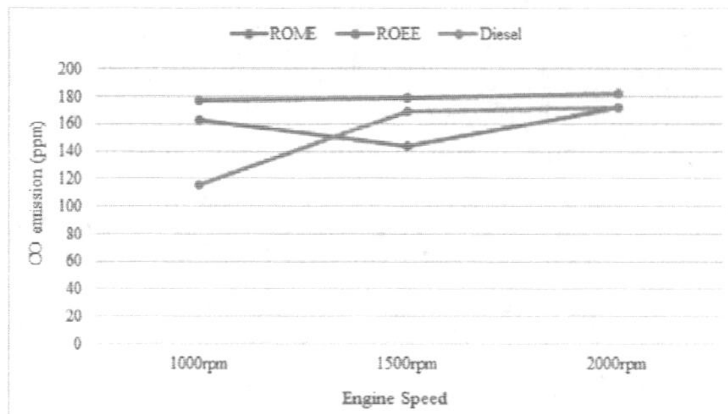

Figure 3. CO emission from diesel engine under different speed.

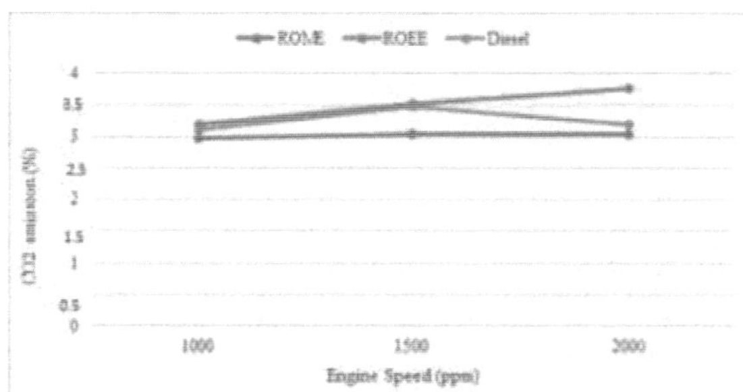

Figure 4. CO_2 emission from diesel engine under different speed.

4. Conclusion

This study has been performed on the production of biodiesel fuel by transesterification reaction and emission characteristics by engine test. From the results, the optimum reaction conditions for the transesterification of rapeseed oil i.e., alcohol/oil molar ratio 18:1, 1% of potassium hydroxide as catalyst, reaction time 30 minutes, reaction temperature 60°C and stirring speed of 650rpm gave the best results. Fuel properties of normal diesel tested in the same condition is closed to the biodiesel and particularly for those produced by using methanol in the transesterification. Results from engine test discloses that that exhaust emission gas such as nitrogen oxides (NOx) and nitric oxide (NO) are reduced with the increase in engine speed in contrary to carbon oxides CO and carbon dioxide CO_2 which are slightly increased. In comparison with diesel fuel, the two biodiesel types (ROME and ROEE) have higher fuel consumption rate because of their lower heating value and their oxygen content. ROME exhibited low emission gas compared to ROEE and mineral diesel as well.

5. References

[1] Amit S 2012 *Biodiesel Production and Properties*. (UK: Royal Society of Chemistry)
[2] Baiju B, Naik M K and Das L M 2009 A comparative evaluation of compression ignition engine characteristics using methyl and ethyl esters of Karanja oil. *Renewable Energy* **34** pp 1616-1621
[3] Bentley R W 2002 Global oil & Gas depletion: An overview. *Energy Police* **30** pp 189–205

[4] Cardone M, Prati M V, Rocco V, Seggiani M, Senatore A and Vitolo S 2002 Brassica carinata as an alternative oil crop for the production of biodiesel in Italy: Engine performance and regulated and unregulated exhaust emissions. *Environ Sci Technol* **36** pp 4656-4662
[5] Demirbas A 1998 Fuel properties and calculation of higher heating values of vegetable oils. *Fuel* **77** pp 1117–1120
[6] Fontaras G, Karavalakis G, Kousoulidou M, Tzamkiozis T, Ntziachristos L and Bakeas E 2009 Effects of biodiesel on passenger car fuel consumption, regulated and non-regulated pollutant emissions over legislated and real world driving cycles. *Fuel* **88** pp 1608-1617
[7] Gerling P 2007 *Reserves Resources and Availability of Energy*. (Hannover: Federal Institute for Geosciences and Natural Resources)
[8] Hansen A C, Gratton M R and Yuan W 2006 Diesel engine performance and NOx emissions from oxygenated biofuels and blends with diesel fuel. *Trans. ASABE* **49** pp 589-595
[9] John B 2007 *Diesel Fuels Technical Review*. (San Ramon, USA: Chevron Corporation)
[10] Kalligeros S F, Zannikos S, Stournas E, Lios G and Anastopoulos C T 2003 An investigation of using biodiesel/marine diesel blends on the performance of a stationary diesel engine. *Biomass and Bioenergy* **24** pp 141–149
[11] Kent Hoekman S, Amber B, Eric C and Curtis R 2012 Review of biodiesel composition, properties, and specifications. *Renewable and Sustainable Energy Reviews* **16** pp 143–169
[12] Klaus M and Helmut T 2010 *Handbook of Diesel Engines*. (Berlin Heidelberg, Germany: Springer-Verlag)
[13] Krahl J, Knothe G, Munack A, Ruschel Y, Schröder O and Hallier E 2009 Comparison of exhaust emissions and their mutagenicity from the combustion of biodiesel, vegetable oil, gas-to-liquid and petrodiesel fuels. *Fuel* **188** pp 1064-1069
[14] Lapuerta M, Armas O and Herreros J M 2008 Emissions from a diesel–bioethanol blend in an automotive diesel engine. *Fuel* **1** pp 25-31
[15] Mert G and Atilla B 2015 Density, flash point and heating value variations of corn oil biodiesel–diesel fuel blends. *Fuel Processing Technology* **134** pp 456-464
[16] Mushtaq A, Mir A K, Muhammad Z and Shazia S 2013 *Practical Handbook on Biodiesel Production and Properties*. (Boca Raton, USA: CRC Press Taylor & Francis Group)
[17] Ozsezen A N, Canakci M, Turkcan A and Sayin C 2009 Performance and combustion characteristics of a DI diesel engine fueled with waste palm oil and canola oil methyl esters. *Fuel* **88** pp 629-636
[18] Rafael L and Juan A M 2012 *Advances in Biodiesel Production Processes and Technologies*. (UK: Woodhead Publishing Limited)
[19] Ramadhas A S, Muraleedharan C and Jayaraj S 2005 Performance and emission evaluation of a diesel engine fueled with methyl esters of rubber seed oil. *Renewable Energy* **30** pp 1789-1800
[20] Usta N 2005 An experimental study on performance and exhaust emissions of a diesel engine fueled with tobacco seed oil methyl ester. *Energy Conversion and Management* **46** pp 2373–2386

IOP Conf. Series: Materials Science and Engineering 1175 (2021) 012002 doi:10.1088/1757-899X/1175/1/012002

Seakeeping and resistance analysis of 1200 GT passenger ship fitted with NACA 4412 stern foil using CFD method

I P S Asmara[1], Sumardiono[2] and M Adam[3]
[1,2,3]Shipbuilding Engineering Department, Politeknik Perkapalan Negeri Surabaya

Email: putusindhu@ppns.ac.id

Abstract. Seakeeping performance is one of the requirements to ensure the convenience and safety of life at sea, especially for passenger ships. In this paper, the result of computational fluid dynamics (CFD) analyses concerning the effect of NACA 4412 stern foil to the improvement of seakeeping quality and resistance of 1200 GT passenger ship are presented. The fitting of stern foil has become a popular method for the improvement of ship design. Stern foil is hydrofoil attached to the transom of the vessel located at the stern part below the waterline. The analysis was developed on the conditions of regular and irregular head waves. The CFD computation on 8 variations of the stern foil position and attack angle has been carried out by using Numeca Fine/Marine. The study confirms that the ship resistance decrease by 3.6 % at the service speed of the ship. The improvement of heaving and pitching amplitudes, as well as heaving acceleration, are significant for the hull vane located at 50% of displacement draft and 1% of displacement length with the attack angle of -4°.

1. Introduction

Ship motion response to regular and irregular waves can be determined from motion analysis, namely maneuvering and seakeeping analysis. The maneuvering usually only considers the 3 degrees of freedom of ship motions in the horizontal plane area including surging, swaying and yawing. Seakeeping analysis concern about the 3 other motions including heaving, pitching and rolling. External excitation such as current, wind, and wave cause additional resistance and the 6 degrees of freedom of ship motion, as seen in Fig.1. The excessive amplitude and acceleration of the motion will cause motion sickness symptoms. The naval architect should consider the ability of the ship to be operated on various sea conditions by complying the requirement of seakeeping criteria. This paper provides the improvement seakeeping performance of 1200 GT passenger ship equipped with NACA 4412 stern foil.

The development of maritime technologies and its support, especially in ship designs has been achieved by the invention of Hull Vane®. It has been invented by Van Oossanen [1]. It is aimed to reduce ship motion and increase performance. Hull vane is a fixed hydrofoil located below the waterline that attached to the transom or stern of the vessel. The research shows there are 4 interrelated effects can be found on the hull vane, a thrust force, a trim correction, the reduction of waves, and the reduction of motions in waves. Based on basic the foil theory, a foil will generate decomposed force vectors, chordwise force and neutral force [2]. Those force will influence on ship motion.

Content from this work may be used under the terms of the Creative Commons Attribution 3.0 licence. Any further distribution of this work must maintain attribution to the author(s) and the title of the work, journal citation and DOI.
Published under licence by IOP Publishing Ltd

Hull Vane® is one of the appendages fitted on the stern which has the potential to increase the ship's resistance. Accordingly, its proper position and attack angle is necessary to be found out in order to ensure that the resistance of ship fitted with hull vane will improve the ship propulsion efficiency. The other study shows that hull vane can lead to a significant reduction in ship resistance [3].

2. Literature Review

Seakeeping performance of the vessel is depent on heading angle of wave, the wave amplitude, and wave spectrum. The ship motions will cause the level of motion sickness incidence and the ship operability. The definition of seakeeping, respon amplitude operator, wave spectrum, motion sickness incidence and operabilit of ship is described as below.

2.1. Seakeeping

The motions of ship considered in seakeeping analysis is divided into 3 motions. i.e. heaving, pitching, and rolling. The pitching response is more significant than the rolling response by considering that the longitudinal distance of the stern foil to the center line of the motion, the moment lever is significantly longer than that of the pitching motion. Figure 1 represents that heaving is the motions of a ship parallel to the vertical direction, Z axis. The heaving is the oscillation motion leading to vertically ship motion, up and down movement. Pitching is the oscillation motions of the ship rotating according to the lateral direction, the Y-axis. The pitching motion represented by the change of ship's trim to the bow and stern. The pattern of the motions depends on the inertial load of the ship and hydrostatic load of the wave. The seakeeping criteria are presented in Table 1.

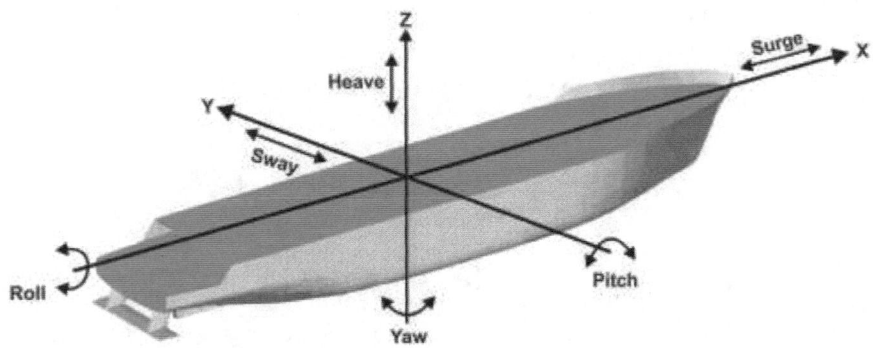

Figure 1. Six Degrees of Ship Motions.

Table 1. Seakeeping Criteria.

No.	General Criteria
1.	RMS Pitch ≤ 1.5° [8]
2.	RMS Heave Acceleration ≤ 0.1g (normal work for the crew) [9]
3.	3° single amplitude average pitch [10]
4.	Significant heave acceleration ≤0.4g (no people working on deck) [10]
5	Significant heave acceleration ≤0.2g (people working on deck) [10]

IOP Conf. Series: Materials Science and Engineering 1175 (2021) 012002 doi:10.1088/1757-899X/1175/1/012002

2.2. Respon Amplitude Operator (RAO)

The motion of a floating vessel due to a passing hydrodynamic regular wave is called the RAO (Response Amplitude Operator), where RAO is the ratio between ship's amplitude of motions (either translation or rotation) to the amplitude of the wave at a particular frequency. RAO or often referred to as the Transfer Function because it is a instrument to transfer the external load (wave) in response to a structure. The general form of the RAO equation in the frequency function is as follows [4]:

$$\text{Response}\ (\omega) = (\text{RAO})\eta(\omega) \tag{1}$$

where, η = wave amplitude, m, ft

The RAO for translational motion is a direct comparison between the amplitude of the ship's motions and the wave amplitude both in length units as shown by (2).

$$RAO = \frac{Z_0}{\zeta_0}\ (\text{m/m}) \tag{2}$$

RAO for rotational motion is the ratio between the amplitude of the rotational motion, in radians, to the wave inclination which is the multiplication of the wave number, $kw = \omega^2 / g$ with wave amplitude as shown by (3).

$$RAO = \frac{\theta_0}{kw\zeta_0} = \frac{\theta_0}{(\frac{\omega^2}{g})\zeta_c}\ (\text{rad/rad}) \tag{3}$$

Unfortunately, RAO can't describe the ship's response to the real state of the ocean because the waves are irregular. Response of ship motions to random wave can be described with response spectrum. The response spectrum is obtained by multiplying the wave spectrum, $S\zeta$ with RAO2 as seen in (4).

$$S_{\zeta r}\ (\omega) = RAO^2 \times S_{\zeta}\ (\omega) \tag{4}$$

2.3. Wave Spectrum

The JONSWAP wave spectrum [3] is used in this analysis. This spectrum is a refinement of the Pierson-Moscowits spectrum. The North Sea has extreme environmental conditions and is bordered by islands and continents which result in fetch in this area quite short but has large waves. So the Pierson-Moscowits equation is changed in the form of equation (5)

$$S(\omega) = \alpha g^2 \omega^{-5} \exp\left[-1.25\left(\frac{\omega}{\omega_0}\right)^{-4}\right] \gamma^{\exp\left[\frac{(\omega_0-\omega)}{2\tau 2\omega_0^2}\right]} \tag{5}$$

where,

γ is peakedness parameter	= 3.3
τ is shape parameter	= 0.07, if $\omega \le \omega z$
	= 0.09, if $\omega > \omega z$
α	= $0.076\ (x_0)^{-0.22}$
	= 0.00819 (when x is unknown)
ω_0	= $2\pi\ (g / U\omega_0)\ (x_0)^{-0.33}$
x_0	= $gx/U\omega^2$
ω_0^2	= $0.161g/HS$

3. Methodology

The CFD analysis is handled by using Numeca Fine/Marine software. The related variables of wave height and period for the motion computation are derived from the data on the same condition of Java Sea shipping route. The results are ship response amplitude of heaving and pitching as well as the heaving acceleration and ship's resistance. Seakeeping analysis is performed in this study by comparing the motion of the ship with hull vane and without the hull vane in regular waves. In addition, the seakeeping analysis of the ship equipped with the stern foil in irregular waves is developed to predict the probability of the ship motion not complying with the seakeeping criteria.
The main dimension of 1200 GT passenger ship analyzed in this study is shown in Table 2 as below. The ship's model is developed in Maxsurf Modeler. It has been verified by using key plan data and stability booklet. The precision of the model is 99.02% of displacement, 99.56% of the coefficient block, and 99.69% of the longitudinal center of buoyancy.

Table 2. Ship's Main Dimension.

No	Dimensions	Passenger Ship
1	Length overall (LOA), meters	62.8
2	Length between perpendiculars (Lpp), meters	57.36
3	Breadth (B), meters	12
4	Height (H), meters	4
5	Draught (T), meters	2.7
6	Service speed (Vs), knots	12
7	Displacement (Δ), tons	1318
8	Gross Tonnage (GT), tons	1200

The stern foil models are designed referring to an embodiment of aft foil patented by Van Oossanen [1]. There are 8 models of the foil differing by foil's reference point position and angle, horizontal position (x-axis), and vertical position (y-axis). The leading edge of foil is used as a reference point. The horizontal position is measured from the stern waterline intersection to the leading edge. The vertical position is measured from keel to the leading edge. The variation is presented in Table 3 as seen in Figure 2 and Figure 3.

Table 3. Foil Position and Angle.

No.	Models	X-axis (m)	Y-axis (m)	Foil Angle (degree)
1	1A	0	1.030	0
2	1B	0	1.538	0
3	1C	0.588	1.030	0
4	1D	0.588	1.538	0
5	2A	0	1.030	-4
6	2B	0	1.538	-4
7	2C	0.588	1.030	-4
8	2D	0.588	1.538	-4

IOP Conf. Series: Materials Science and Engineering 1175 (2021) 012002 doi:10.1088/1757-899X/1175/1/012002

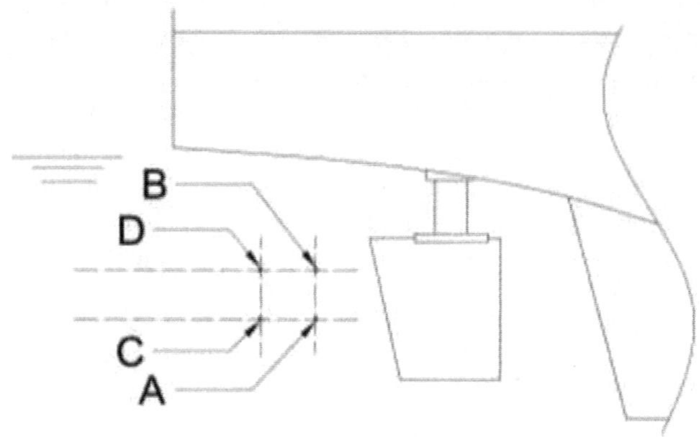

Figure 2. Six Degrees of Ship Motions.

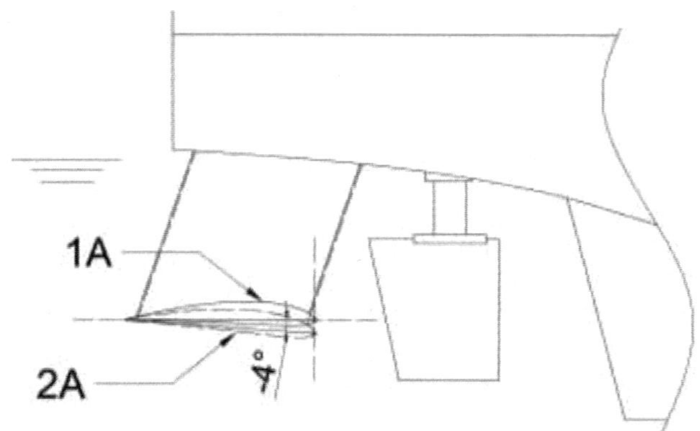

Figure 3. Six Degrees of Ship Motions.

The meshing model of the ship fitted with the NACA 4412 stern foil is shown in Figure 4. The number of cell meshing of the model is 667461 cells selected by comparing the result on the trend of the motions by the number of cell meshing as shown in Figure 5.

Figure 4. Mesh model of the ship Fitted with the Stern Foil.

IOP Conf. Series: Materials Science and Engineering 1175 (2021) 012002 doi:10.1088/1757-899X/1175/1/012002

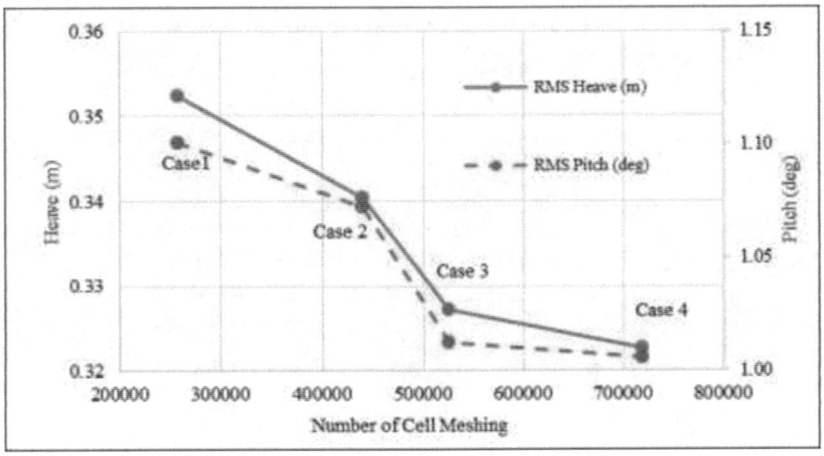

Figure 5. RMS of Ship Motions by Cell Meshing.

Firstly, seakeeping analysis of root mean square (RMS) pitch, RMS heave, and RMS heave acceleration on a regular wave with height significant (Hs) Of 1.8 meters at the Java Sea and zero up crossing wave period is 7s. The RMS value is calculated after the ship has reached steady motion. Secondly, JONSWAP wave spectrum is chosen to evaluate the acceptance of the ship's motion according to the seakeeping criteria and to calculate the probability of the motions out of the criteria. The criteria are presented in Table 3. Finally, the mean value of ship resistance on the regular wave at the medium and full speed is simulated. The position of the ship in the computational domain is shown in Figure 6. The distances are 1.0 length overall (LOA) to the inlet, 3.0 LOA to the outlet, 1.5 LOA the bottom, 0.5 LOA to the top, and 1.5 LOA to both sides.

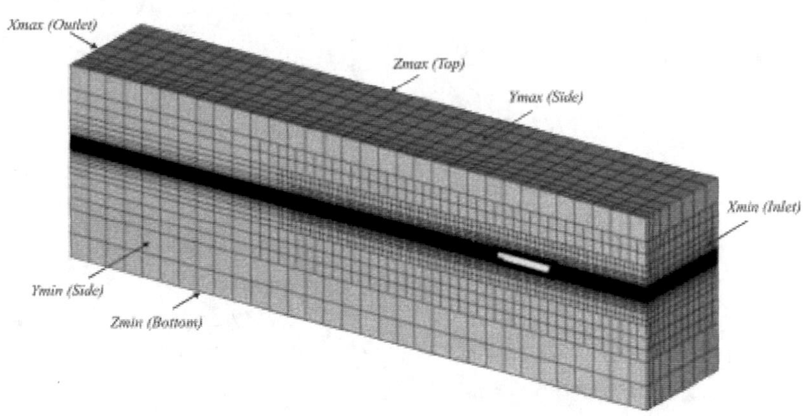

Figure 6. Ship Position in the Computational Domain.

4. Results and Discussion

4.1. Motions on Regular Wave

The root mean square (RMS) motions at the condition of the regular head sea is shown in Fig. 4. The responses of 8 models compared to those of the model without Hull Vane® (NOHV) are presented in Table 3. The most reduction on the root means square (RMS) motions happen on model 2C. The RMS pitch is reduced by -10.75%, the RMS heave by -6.96% and the RMS heave acceleration by -7.32%, as presented by Figure 7.

IOP Conf. Series: Materials Science and Engineering 1175 (2021) 012002 doi:10.1088/1757-899X/1175/1/012002

Table 3. Comparison Of Motions.

Model	RMS Pitch (deg)	RMS Heave (m)	RMS Heave Acceleration (m/s^2)
NOHV	1.1078	0.3411	0.6823
1A	1.0118	0.3298	0.6716
1B	1.0275	0.3295	0.6689
1C	0.9936	0.3244	0.6630
1D	1.0067	0.3253	0.6625
2A	1.0127	0.3272	0.6558
2B	1.0226	0.3288	0.6582
2C	0.9887	0.3174	0.6323
2D	0.9987	0.3280	0.6558

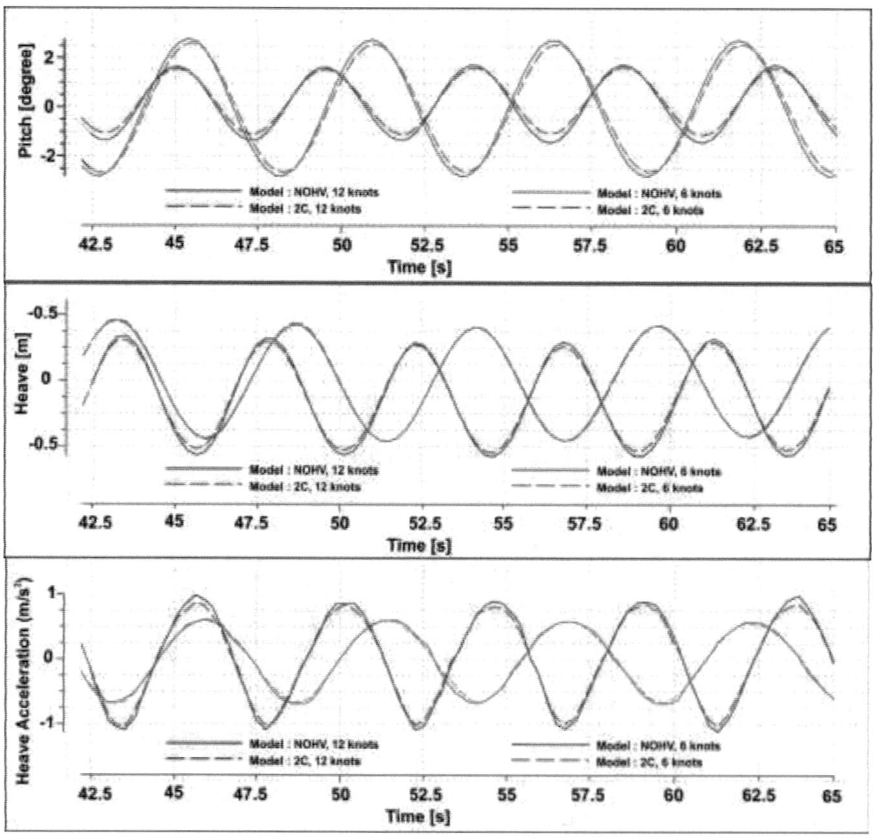

Figure 7. Pitch, Heave, and Heave Acceleration Recponses to Regular Wave.

4.2. Resistance

The reduction of ship resistance fitted with the foil at the full speed of 12 knots is shown in Figure 8. The RMS resistance of ship using NOHV model is -68,157.41 N and 2C model is -65,672.91 N. The negative value of the X force shown in Figure 7 means the force against ship direction. The turbulent viscosity of the ship with stern foil 2C is presented in Figure 9.

IOP Conf. Series: Materials Science and Engineering 1175 (2021) 012002 doi:10.1088/1757-899X/1175/1/012002

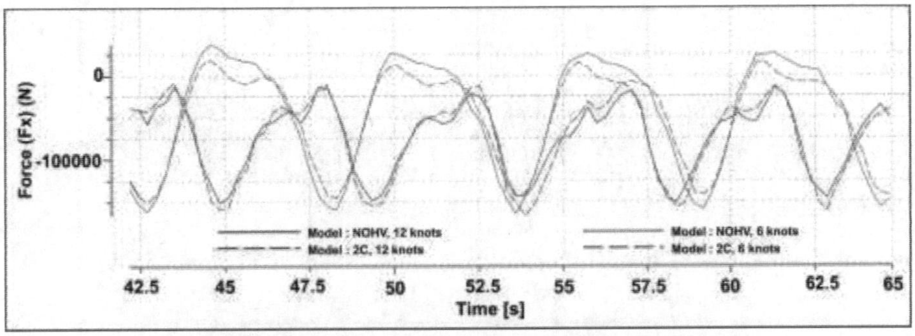

Figure 8. Ship Resistance at Medium and Full Speed, 6 kts and 12 kts.

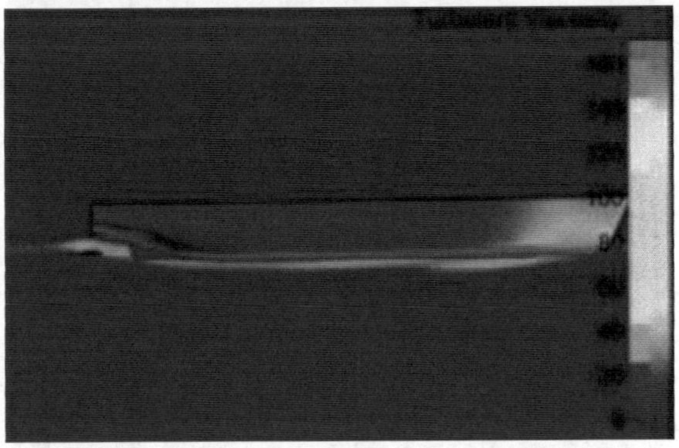

Figure 9. Turbulent Viscosity of Model 2C.

4.3. Seakeeping Performance

Computation using an irregular wave of JONSWAP spectrum with height significant, Hs of 1.8m, period, T of 7s, and peakedness parameter, $\gamma = 2.5$ for model 2C results in RMS pitch of 0.8220 and RMS heave acceleration of 0.0416g. The response motions are presented in Figure 10. The probability of the motions comply with the criteria are presented in Table 7.

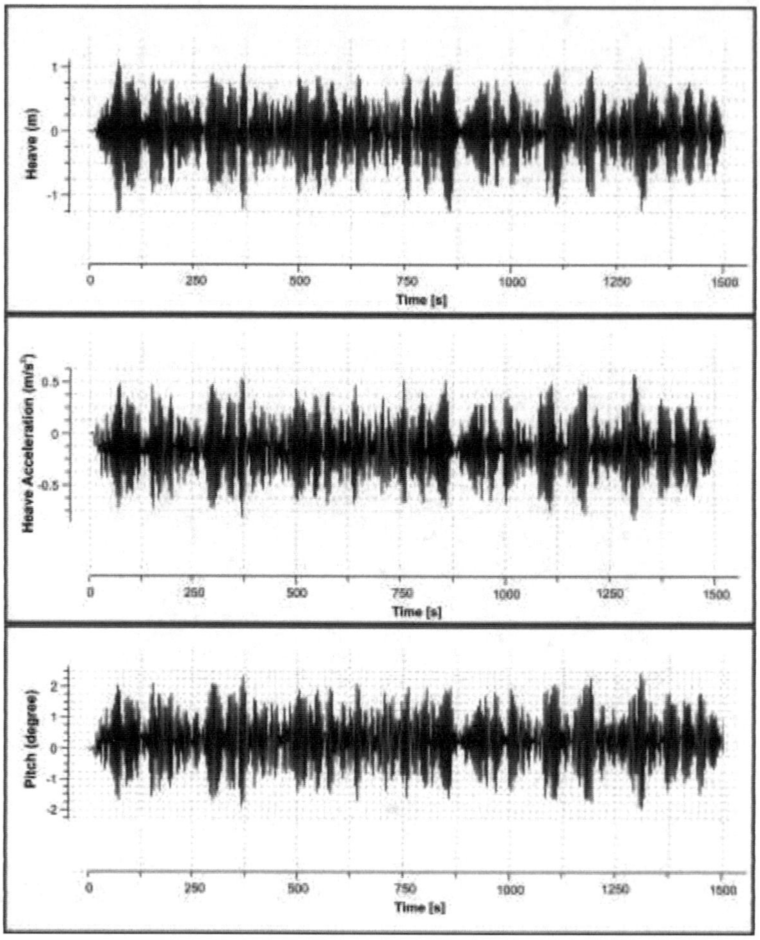

Figure 10. Pitch, Heave, and Heave Acceleration Responses to Irregular Wave.

Table 7. Probability of Complying The Criteria.

No	Criteria	Probability
1	RMS Pitch ≤ 1.5°	76%
2	RMS Heave Acceleration ≤ 0.1g	93.67%
3	3° single amplitude average pitch	100%
4	Significant heave acceleration ≤0.4g	100%
5	Significant heave acceleration ≤0.2g	100%

5. Conclusion

Analysis of resistance and seakeeping of 1200 GT passenger ship using CFD computation based on responses to regular and irregular wave is concluded as below.

a) NACA 4412 stern foil fitted according to model 2C can reduce the resistance by 3.6% at the service speed.

b) The RMS pitch, heave and heave acceleration decrease up to 10.75%, 6.96%, and 7.325, respectively.

c) The probability of the ship fitted with the foil to comply with the seakeeping criteria is 76%.

6. References

[1] Oossanen V 2016 *Vessel Comprising an Aft Foil Oriented to Provide a Forwardly Directed Component of Lift Force.* International Application Published Under The Patent Cooperation Treaty No WO 2016/010423 Al

[2] Hagemeister N et al 2017 A comparison between four alternatives for A 61m offshore patrol vessels. *Proceedings of Fast Sea Transportation (FAST)* (27-29 September Nantes – France)

[3] Karthik A V 2013 *CFD Analysis of Resistance Characteristics of High-Speed Displacement Hull Forms fitted with Hull Vane.* (Thesis Master of Science in Ocean Engineering Florida Institute of Technology)

[4] Bhattacharyya R 1972 *Dynamics of Marine Vehicle.* (Annapolis Maryland: John Willey & Sons New York)

[5] Hidayat K S A and Riaydi S 2017 Effects of the application of A stern foil on ship resisiance: A case study of an orela crew boat. *International Journal of Technology* **7** pp 1266-1275

[6] Uithof B B K et al 2016 Hull vane on Holland-class OPVs - A CFD analysis of the effects on seakeeping. *13th International Naval Engineering Conference and Exhibition (INEC)* (Bristol)

[7] Steen E B S 2013 The effect of a fixed foil on ship propulsion and motions. *Third International Symposium on Marine Propulsors (SMP)* (Launceston Tasmania: Australia)

[8] Cummins W E et al 1989 *Principles of Naval Architecture Volume III - Motions in Waves and Controllability.* (The Society of Naval Architects and Marine Engineers Jersey City: NJ)

[9] Nordforsk 1987 *The Nordic Coorperative Project Seakeeping Performance of Ship Assessment of a Ship's Performace in a Seaway Marintek.* (Trondheim)

[10] Olson S 1978 An evaluation of the seakeeping qualities of naval combatant. *Naval Engineers Journal (ASNE)* **90** 1 pp 23-40

7. Acknowledgement

The authors deliver acknowledgment to the Center of Research and Public Services of Politeknik Perkapalan Negeri Surabaya for providing the facility and supporting fund of this study and publication.

IOP Conf. Series: Materials Science and Engineering 1175 (2021) 012003 doi:10.1088/1757-899X/1175/1/012003

The modelling of particulate matter concentration using mass balance model in surface coating activities at finishing room of shipbuilding process

I Santiasih[1]

[1]Department of Safety Engineering, Politeknik Perkapalan Negeri Surabaya, Keputih, Sukolilo, 60111 Surabaya, Indonesia

E-mail: indri.santiasih@ppns.ac.id

Abstract. The surface coating process of shipbuilding produces pollutants in the form of particulate matter (PM_{10} and $PM_{2,5}$), potentially generating environmental pollution and respiratory problems in workers. The concentration of indoor particulate matter (PM) can be predicted using mass balance model. This study aims to prove that mass balance can be employed to estimate the concentration of PM. Mass balance is calculated from the concentration of pollutants, inserted into indoor, indoor emission rate, airflow rate and pollutant removal rate. The model estimation results show that PM_{10} concentrations on the first day, second day, third day and fourth day were 1,22 mg/m^3, 1,52 mg/m^3 ,1,84 mg/m^3 and 1,80 mg/m^3. While the $PM_{2,5}$ concentration on the first day, second day, third day and fourth day was 2,67 mg/m^3, 3,03 mg/m^3, 2,99 mg/m^3 and 3,15 mg/m^3. The RMSE values on day 1 to day 4 for PM_{10} were 0.679412, 0.565597, 0.682349, 0,546626, while for $PM_{2.5}$ are 0,755513, 0,980255, 0,975141 and 1,108106. Giving evidence that mass balance model was reliable that proved by the lower value of RMSE which is generally less than one. It proves that the concentration of PM (PM_{10} dan $PM_{2,5}$) is close to actual value of PM concentration from measurement results.

1. Introduction

The surface coating is one of the production activities of shipyard companies in East Java. The process carried out in the finishing room includes the activities of giving fillers (putty), rubbing (blasting) and basic painting. The process of painting executes with a spray gun is more dangerous than paintbrushes due to the small particles can be released into the environment. The application of spray painting operation generates the ultrafine particles (Santiasih and Hermana 2017) that discharge to the environment and then fly in the air, thus it can contaminate the environment and endanger the human health (He *et al.* 2019). Spray paints convert substances into aerosols and /or particulate matter (PM), which is a group of ultrafine particles in the form of liquid and/or solid substances. PMs are obtained from spraying mechanisms, nebulization and bubble producing (Santiasih, Titah, and Hermana 2020). Particulate matter with a small size (ultrafine PM) will be easily inhaled during inhalation exposure, thus it is potential exposure especially to the respiratory system. The empirical relationship between exposure to particulate matter with enzymatic effects on mammals (Santiasih *et al.* 2019) and advanced effects on the respiratory system, cardiovascular system and particulate concentrations in ambient air has been conducted by a previous study (USEPA 1997).

Content from this work may be used under the terms of the Creative Commons Attribution 3.0 licence. Any further distribution of this work must maintain attribution to the author(s) and the title of the work, journal citation and DOI.
Published under licence by IOP Publishing Ltd

IOP Conf. Series: Materials Science and Engineering 1175 (2021) 012003 doi:10.1088/1757-899X/1175/1/012003

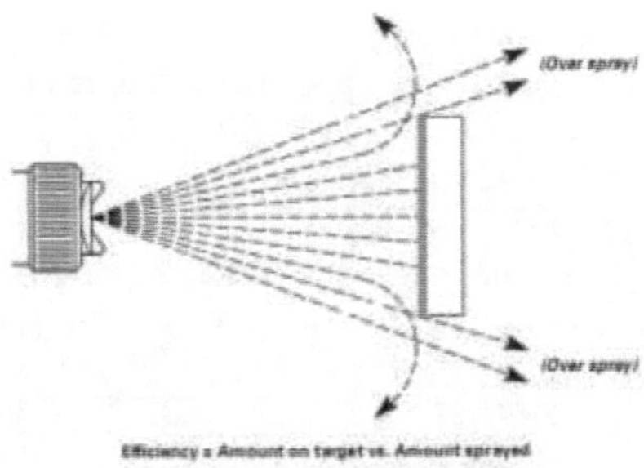

Figure 1. The Illustration of Transfer Efficiency
Source: Graco Inc. (2014).

Spray painting method in the surface coating process which has a low transfer efficiency, thereby releasing more aerosols into the air (overspray) compared to aerosols that attach to the work plane (Figure 1). The process of surface coating produces pollutants, namely secondary PM with a very small size (ultrafine PM). Secondary particles are generated as a result of chemical reactions from precursors substances of volatile organic compounds (VOC) such as sulphate, nitrate, Sulphur dioxide, nitrogen dioxide and ammonia (Lazaridis *et al.* 2015a). Volatile organic compounds (VOCs) from painting solvents are one of the most important sources of pollutant outputs for the shipbuilding and ship repair industry (Celebi and Vardar 2008a). Impacts caused by PM are exposure to smoke or fog, contamination of buildings and other materials, corrosion and material damage and also initiate local weather changes. Besides, particulates also cause interference with humans, animals and plant growth. Some scientist has been investigated health impact of PM in the respiratory system, cardiovascular system, and cellular effect (Franck *et al.* 2011; Karottki *et al.* 2014; Orona *et al.* 2014).

Indoor air quality (IAQ) is currently attracting the attention of many experts, especially environmental engineers and industrial hygienists because the importance of the indoor environment is related to indoor exposure to workers. The exposure of indoor PM has to obtain serious attention, considering that humans spend most of their time (80%) indoors whether in homes, offices, vehicles, workplaces, laboratories (Myers and Maynard 2005). Cooper dan Alley (2011) investigated that the concentration of indoor air pollutants can be predicted using a simple mathematical model, with the key variables being the emission rate and ventilation rate, such as the mass balance model. In a typical indoor air mass balance model, the assumption is generally formulated that the indoor air is well- mixed (Wadden and Scheff 1983). The mass balance equation is based theoretically on the physical law of mass conservation, to calculate the particle concentrations in indoor regarding all parameters such as particle outdoor concentration that inserted into indoor, initial indoor concentration, indoor emission rate, sink removal and also considering indoor airflow rate (Cooper and Alley 2011).

It is essentials to investigate the estimation of indoor PM (PM_{10} and $PM_{2,5}$) using the mass balance model, to examine the reliability of the mass balance model in predicting the concentration of indoor PM. The results of this study can be an alternative method to determine the concentration of indoor PM if a direct measurement is difficult or not possible to be carried out.

2. Material and Methods

2.1. The Measurement of Particulate Matters (PM)

$PM_{2.5}$ and PM_{10} measurements were conducted using a *High-Volume Air Sampler* (HVAS) in the breathing zone in height from the floor to represent inhaled PMs. The PM air sampling in ambient air (indoor) was carried out when the production process was running normally (no reduction or addition of raw materials, and production time). A sampling of PM_{10} and $PM_{2,5}$ was conducted at 6 points (P) in the finishing room (Figure 2) with 4 repetitions. The measured PM_{10} and $PM_{2.5}$ concentrations were in ambient air indoors, and outdoors.

2.2. The Measurements of Air Flow and Air Velocity

The airflow rate and air velocity were measured at 6 points (P) in the finishing room (Figure 2) by the repetition of 4 times measurement, using the wet bulb globe temperature (WBGT) *instrument Questtemp Heat stress monitor model 32.*

Figure 2. The Layout of Finishing Room.

2.3. The Modelling Concentration of Particulate Matters using the Mass Balance Model

The concentration of indoor air contaminants was predicted using simple mathematical models such as mass balance, with the key variables being the rate of emissions and the rate of ventilation. In some buildings, the airflow rate was very simple, so it could be assumed that the building had a single airflow rate and the also contained perfectly mixed air. The mass balance equation was based theoretically on the physical law of mass conservation, to calculate the particle concentrations in indoor regarding all parameters such as particle outdoor concentration that inserted into indoor, initial indoor concentration, indoor emission rate, sink removal and also considering indoor airflow rate (Cooper and Alley 2011).

$$\frac{VdC_i}{dt} = QC_0 + S - QC_i - kC_iV \quad (1)$$

Where:

V	= room volume, m^3
C_i	= concentration of indoor pollutants, $\mu g/m^3$
C_0	= the concentration of pollutants in the outside air, $\mu g/m^3$
Q	= ventilation rate, m^3/hour
S	= rate of emission source in the room, μg/hour
k	= the constant of removal rate (assumed beforehand), $hour^{-1}$

Thus, Equation 1 could be written as:

$$\frac{dC_i}{dt} + \left(\frac{Q}{V} + k\right)C_i = \frac{Q}{V}C_0 + \frac{S}{V}$$

The value of $(Q/V + k)^{-1}$ was a time characteristic or time constant for this system and was given the symbol τ. The general solution for Equation 1 was:

$$C_i = \left\{C_0 - \tau\left(\frac{Q}{V}C_0 + \frac{S}{V}\right)\right\}\{exp[-t/\tau]\} + \tau\left(\frac{Q}{V}C_0 + \frac{S}{V}\right) \quad (2)$$

Where C_0 was the initial concentration in the room at time 0. The Q / V value included the air exchange rate (hour^{-1}) or could be interpreted as the room volume of the air exchange per hour. The solution for the steady-state case from Equation 2 was:

$$C_{i\,ss} = \tau\left(\frac{Q}{V}C_0 + \frac{S}{V}\right) \quad (3)$$

Based on Equation 3, the mass balance model for indoor air quality was:

$$C_{i\,ss} = \frac{\left(\frac{Q}{V}C_0 + \frac{S}{V}\right)}{\frac{Q}{V} + k} \quad (4)$$

or

$$C_{i\,ss} = \left(\frac{AC_0 + S/V}{A+k}\right)$$

Where A = Q/V air exchange in each hour (Cooper dan Alley, 2002).

Nevers (1995) examined the indoor air quality model that formulated more realistic and more complex

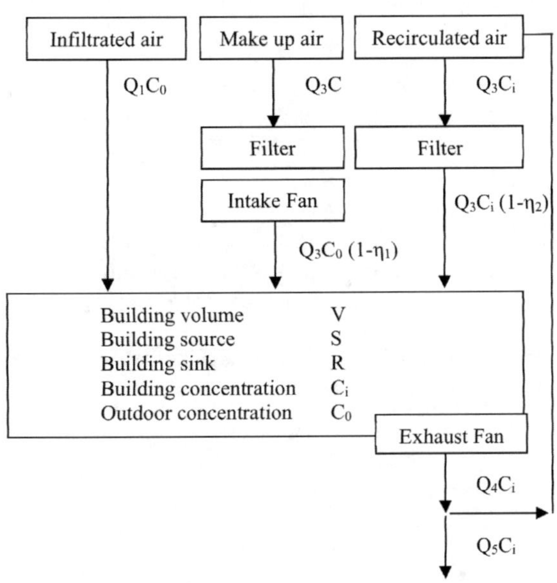

Figure 3. The Flow Chart of Model Constructed with Well-Mixed of Indoor Air (Shair and Heitner 1974).

Figure 3 illustrated that there was a building with air infiltration, ventilation with treatment, and recirculation air with separate treatment. Thus, it could be assumed that the building had perfectly mixed air. This model could be compared with real building conditions, where there were many rooms with simultaneous airflow from one room to another, and had different emission rates and concentrations in several rooms.

The material balance equation for the room in Figure 3 was:

$$V\frac{dCi}{dt} = Q_1C_0 + Q_3C_0\,(1-\eta_1) + Q_4C_i\,(1-\eta_2)] - (Q_2+Q_4+Q_5)C_i + S - R \quad (5)$$

Where:

η_1 dan η_2	= removal efficiency of each filter (%)
Q	= air flow rate (m^3/s)
R	= pollutant removal rate (μg/s)
S	= emission rate (μg/s)
C_0	= outdoor concentration (μg/m^3)
C_i	= indoor concentration (μg/m^3)
t	= time (s)
V	= room volume (m^3)

Figure 3 showed that the filter conducted was a wide-ranging purpose, meaning that this filter could be used not only for one pollutant but also could remove several pollutants.

Nevers (1995) investigated the mass balance model was a simple fixed box model that was applied to buildings with well-mixed of internal air conditions with only one inlet and one outlet, in this case, infiltration and exfiltration. In other words, it illustrated that the airflow into and out of buildings through windows, wall openings and roofs that were driven by wind and /or temperature differences inside and outside the room were not considered. Thus, the treated air (makeup water), recirculated air and air through exhaust were assumed to be zero for this simple box model.

The mass balance of pollutants under steady-state conditions with some of the assumptions was explained as follows:
Outflow pollutant = inbound pollutant flow + emission generated in the room - excluded pollutants. Thus, if this condition was explained in the form of an equation, it could be:

$$Q_2C_i = Q_1C_0 + S - R \quad (6)$$

Since the changes in temperature and humidity of the air that passed through this structure were ignored, the values of Q_2 and Q_1 were approximately equal, so the equation could be simplified to:

$$C_i = C_0 + \frac{S-R}{Q} \quad (7)$$

2.4. The Estimation of Particulate Matter Emission Rate in Surface Coating Process

Particulate emissions were calculated in three basic stages of surface coating, namely the preparation stage, the painting stage and the equipment cleaning stage (Table 1). Table 1 explained some typical emissions produced at each stage (NPI 1999b).

In the process of surface coating, a significant activity producing PM was the process of rubbing and painting. Generally, the painting process was executed by a spray system. This spraying process produced fine and ultrafine dust or aerosols (NPI 1999a).

IOP Conf. Series: Materials Science and Engineering 1175 (2021) 012003 doi:10.1088/1757-899X/1175/1/012003

Table 1. The Raw Materials Used and Emissions Generated in the Surface Coating Process.

Process Stages	Raw Material Used	Emissions Generated (According to the List on the NPI)
Preparation and Scouring	Sandpaper, solvent, detergent, acid and caustic soda.	Particulates, including lead and other metals,), sludge (paint), liquid waste.
Painting	Paint, lacquer, varnish, primers.	Individual and total organic compounds in coatings and thinners.
Equipment Cleaning	Solvents and thinners	Individual and total organic compounds in coatings and thinners, liquid waste.

Source: (NPI 1999a, 1999b)

The estimation of many materials from coatings that evaporated in the air during the polishing process was conservatively around to 10% of polishing and painting (NPI 1999a) using the following Equation 8.

$$PM = A \times OpHrs \times S_A \times Po \quad (8)$$

Where:
PM = the particulate emission rate originating from the scrubbing process (kg/hour)
A = the activity rate of scrubbing material usage (kg/hour)
OpHrs = total hours of the scrubbing process (hour/day)
S_A = the percentage of solid ingredients on the scrubbing material.
Po = the percentage of PM emitted during the scrubbing process.

Table 2. The Emission Factors of PM_{10} and $PM_{2,5}$.

Surface	PM Emission Factors, kg/kg sand	
	PM_{10}	$PM_{2,5}$
Precleaned	0.0099	0.0018
Painted	0.022	0.0011
Oxidized	0.0074	0.0011
Average	0.013	0.013

Source: USEPA (1997)

The emission factors of PM emitted from abrasive materials according to USEPA AP-42 (1999) were 0.0099 kg/kg for PM_{10} and 0,0018 kg/kg for $PM_{2,5}$.

Whereas the typical density and the percentage of solid ingredients used in coatings were as follows in Table 3.

Table 3. The Typical and Percentage of Solid Density in Surface Coating.

Type of Coating	Density		Solids (% in Volume)
	kg/L	lb/gal	
Enamel, air dry	0.91	7.6	39.6
Enamel, baking	1.09	9.1	42.8
Acrylic enamel	1.07	8.9	30.3
Alkyd enamel	0.96	8.0	47.2
Primer surfacer	1.13	9.4	39.0
Primer, epoxy	1.26	10.5	57.2
Varnish, baking	0.79	6.6	35.3
Lacquer, spraying	0.95	7.9	26.1
Vinyl roller coat	0.92	7.7	12.0
Polyurethane	1.10	9.2	31.7
Stain	0.88	7.3	21.6
Sealer	0.84	7.0	11.7
Magnet wire enamel	0.94	7.8	25.0
Paper coating	0.92	7.7	22.0
Fabric coating	0.92	7.7	22.0

Source: USEPA, 1995

The equation applied to estimate the emission rate of PM_{10} dan $PM_{2,5}$ was as follows in Equation 9.

$$PM = \rho_n \text{ x } S_T \text{ x } S_A \text{ x Po} \qquad (9)$$

Where:
ρ_n = mixture density.
S_T = maximum spray gun rate (L/hour).
S_A = the percentage of solid ingredients on the coating layer.
Po = the percentage of PM_{10} or $PM_{2,5}$ during the painting process.

2.5. The Validation of the Mass Balance Model

The performance of the mass balance model was evaluated using the Root Mean Square Error (RMSE), by comparing the concentration of PM measurements with the estimated concentrations using the mass balance model. An evaluation of the model performance was carried out to assess the accuracy of the relative predictions of the model with the results of observations, giving the meaning that a low RMSE closed to real value (Ait-Amir, Pougnet, and El Hami 2015). RMSE values indicated the level of error from the modelling results compared with observation results. The error level of a model was low if it had a total RMSE value closed to zero. Featherstone *et al.* (1997) examined that the RMSE used to measure the deviation between random variables with some standard or accepted value.

$$RMSE = \sqrt{\frac{\sum_1^n((x_i-a_i)^2}{N}} \qquad (10)$$

Where x_i was a random variable of N data set, and a_i was the *accepted value*. This criterion in Equation 10 measured the mean square error committed in the test data. In other words, a low MSE value meant that the predicted values are close to the real values (Ait-Amir *et al.* 2015).

3. Results and Discussion

The process of the surface coating produced pollutants, namely secondary PM with a very small in size (ultrafine PM) (Lazaridis *et al.* 2015a). In the finishing room, there were three stages of the process of coating (surface coating), namely the stage of preparation (surface preparation), the stage of painting (painting) and the final stage (drying and storage of finished products). In general, the process can be explained in Figure 4.

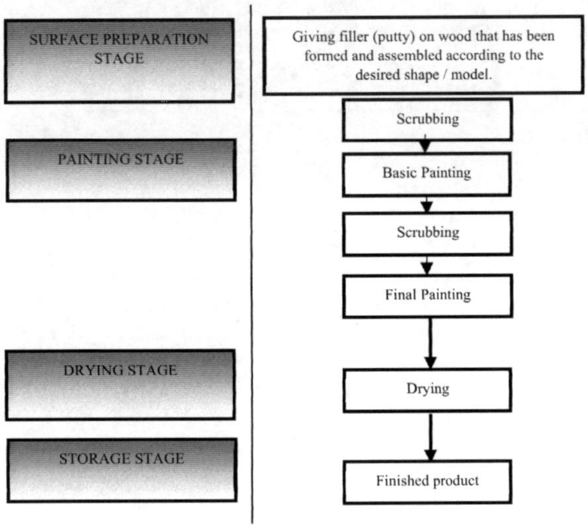

Figure 4. The Flow Diagram of Surface Coating Process.

3.1. The Measurement of Particulate Matter Concentration

The Measurement of particulate concentrations was carried out at 6 points in the finishing room (Figure 2) with 4 repetitions. USEPA (2004) investigated that the mass balance estimation method was suitable for estimating the emissions from sources that produced evaporated pollutants, assuming 100% of the solvent was released into the air through evaporation.

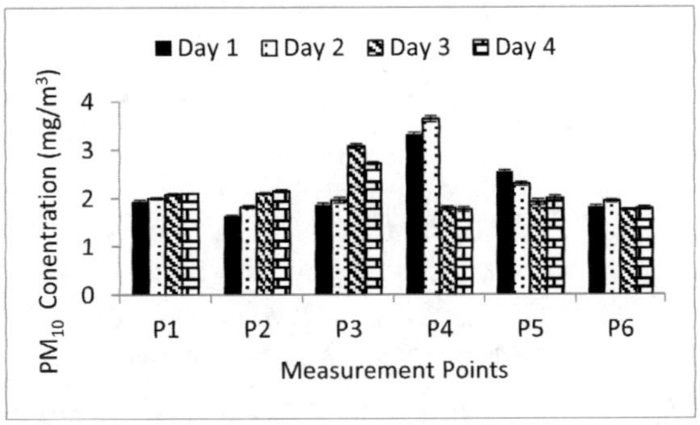

Figure 5. The PM_{10} Concentration of PM Measurements. Data presented in mean and standard deviation.

Figure 5 illustrates the largest of PM_{10} mass concentration was point 4 in day 2 of 3.64 ± 0.06 mg/m^3. The largest mass concentration of $PM_{2.5}$ was also in point 4 in day 2 of 2.36 ± 0.06 mg/m^3 (Figure 6). Point 3 and point 4 were the area of basic painting. Painting activities generated fine PM that was produced secondary PM by a chemical reaction between volatile organic compound (VOC) and ozone and ambient air (Celebi and Vardar 2008b; Lazaridis *et al.* 2015b).

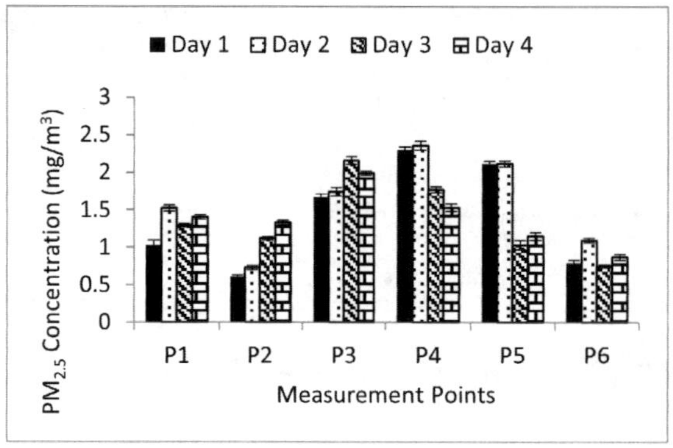

Figure 6. The $PM_{2.5}$ Concentration of PM Measurements.
Data presented in mean and standard deviation.

The process sequence of the surface coating after the basic painting was blasting/scrubbing that was carried out at area point 3 and 4 as well. The scrubbing process produced coarse PM. The coarse PM had a significant influence on the mas PM, unlike fine PM that had no effects on the mass PM (Kelly and Fussell 2012). Giving evidence why point 3 and 4 illustrated the largest mass-PM both PM_{10} and $PM_{2.5}$.

3.2. The Measurement of Air Flow and Air Velocity

The surface coating process carried out in the finishing room was surface preparation and basic painting. The room had an area of 22.66 m^2, not completely closed because it was only given a dividing partition with another room, with a height of 5.4 m. There was no filter or blower in this room, only employed openings of 13.2 m^2. Given the absence of mechanical ventilation in this room, the airflow was low, as listed in Figure 7.

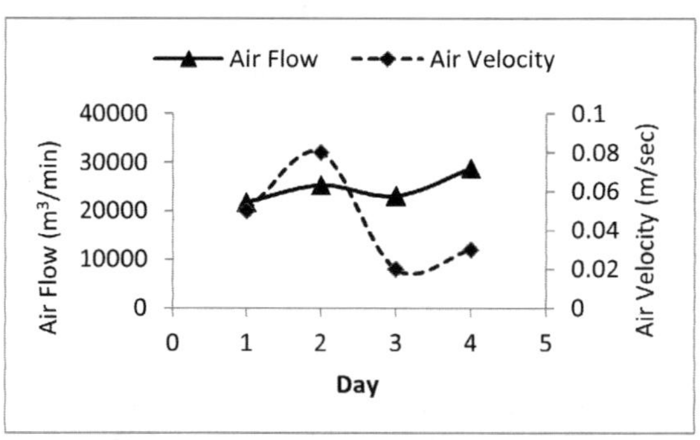

Figure 7. The Airflow and Air Velocity of Measurement.
Data presented in Mean of Each Day Measurements.

Figure 7 illustrates the air velocity was very low because there were only some fans in the large room. While the airflow was high enough, thus it fulfilled the Standard Nasional Indonesia (SNI 2001) threshold value, that the minimum value of Air Change per Hour (ACH) was 6.

3.3. The Estimation of Emission Rate in Surface Coating Process

The surface coating process included three stages as described in Figure 4, thus, the emission rate was determined in every each of these stages.

3.3.1. The Estimation Emission Rate in The Surface Preparation.

The preparation stage (surface preparation) was the initial stage before the wood/raw material was painted. The purpose of this step was to close the wooden pores to make them appeared smoother. The materials applied at this stage were fillers and thinners. After giving the filler (putty), the wood was then scrubbed using sequential sandpaper starting from number 180, number 240 and finally number 400. Scrubbing in the process of surface coating aimed to roughen the material surface as preparation for painting and coating (Lee, 2009). The scrubbing process was conducted 2 times. The first scrubbing was executed after giving filler (putty) and the second scrubbing was carried out after basic painting. The emission rate resulted from the scrubbing process could be calculated using Equation 8. According to USEPA (1997), S_A values for primary surfaces and the emission factors of PM that were released during surface preparation referred to Table 2 (USEPA 1995).

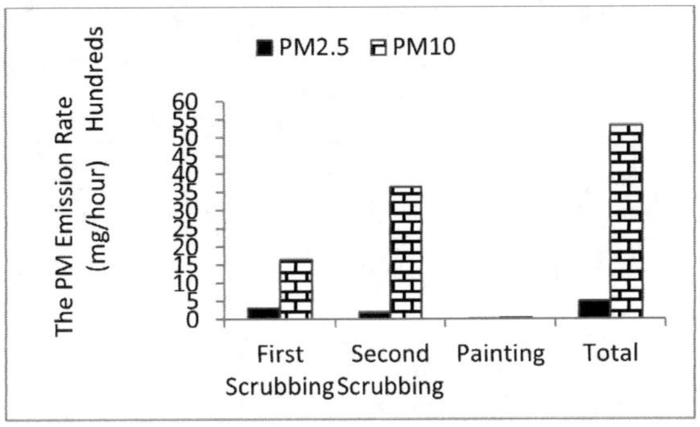

Figure 8. Particulate Matter Emission Rate on Surface Coating.

The emissions rate resulting from the scrubbing activity illustrated in Table 4.

Table 4. The Model Estimation of PM Emission Rate in the Surface Coating Activity.

Surface Coating Activity	Emission Rate (mg/hour)	
	PM_{10}	$PM_{2.5}$
Scrubbing		
First scrubbing	$16,41.10^2$	$2,98.10^2$
Second scrubbing	$36,47.10^2$	$1,82.10^2$
Painting		
Basic painting	$36,47.10^2$	$1,82.10^2$
Total	5325,29	487,24

3.3.2. The Estimation of Emission Rate in The Painting Process.

The painting process was carried out by spraying using a spray gun. This process included basic painting and final painting. The painting process executed in the finishing room was basic painting. Whereas the final painting was conducted in a separate room. The composition of the paint layer applied in the basic painting process was colouring (stain), hardener and thinner. While the spray gun used was a high-volume low pressure (HVLP) type with a transfer efficiency of 65%. The particulate emission rate that results from the painting process according to Equation 9. The mixture density of the painting solution was 1,1247 kg/m^3 and the maximum spray gun rate was 2.4 L/ hour. USEPA (1995) examined that S_A value for epoxy primers referred to Table 3. According to USEPA (1995), the Po value for PM_{10} was 0.069 and 0.0134 for $PM_{2.5}$. Thus, the emission rate from the base painting activity for PM_{10} was 37.29 mg/hour, while for $PM_{2,5}$ was 7.24 mg/hour (Figure 8). The total emission rate generated in the coating process (surface coating) in the finishing room shows in Table 4.

3.4. The Mass Balance Model

In the finishing room, there was no exhaust fan. It was assumed there was no pollutant removal process in the finishing room (R = 0), thus, Equation 7 for the estimated $PM_{2,5}$ concentration on the first day became:

$$C_{PM} = C_0 + \frac{S}{Q} \tag{11}$$

Table 5. The Estimation of PM Concentration.

Days	Concentration C_0 (mg/m^3)				Concentration Estimation Results (mg/m^3)	
	PM_{10}		$PM_{2,5}$		PM_{10}	$PM_{2,5}$
	Out1	Out2	Out1	Out2		
First	0,61	0,61	1,03	1,64	1,22	2,67
Second	0,67	0,85	1,21	1,82	1,52	3,03
Third	0,79	1,05	1,25	1,74	1,84	2,99
Fourth	0,80	1,00	1,40	1,75	1,80	3,15

Out1 = outdoor 1; Out2 = outdoor 2

Cooper and Alley (2002) investigated that the concentration of indoor air contaminants could be predicted using a simple mathematical model, with the key variables being the emission rate and ventilation rate. The concentration of PM_{10} and $PM_{2,5}$ was estimated for 4 days as illustrated in Table 5. The results show that the estimated PM_{10} and $PM_{2,5}$ concentrations had the accuracy of up to $\frac{1}{1000000}$ (for example, the estimated $PM_{2,5}$ results on the first day was 2,670373). However, considering the accuracy of the measuring instrument just only reached $\frac{1}{100}$ (two digits after the decimal point), the results were depicted two digits after the decimal point. Regarded on the estimated concentration of PM_{10} and $PM_{2,5}$ on 4 days of sampling, it could be calculated the estimated concentration of PM_{10} and $PM_{2,5}$ that left in the room after the work process was finished (after 7 hours of work) using Equation 7. There was no air treated through the filter (R = 0), no air recirculated (Q3 = 0) and there was no exhaust fan (Q4 and Q5 = 0), thus, the equation was only influenced by infiltrated air into the room. So, Equation 7 modified as followed in Equation 12.

$$\begin{aligned} V\frac{dCi}{dt} &= Q_1C_0 - Q_2C_i \\ \frac{dCi}{C_0-C_i} &= \frac{Q}{V}\,dt \\ \ln\frac{C_o-C_i}{(C_o-C_i)awal} &= -\frac{Q}{V}\,\Delta t \end{aligned} \tag{12}$$

The estimation of PM concentration was determined using Equation 12, and the results were depicted in Table 6.

Table 6. The Estimation of PM Concentration that was Left Behind in the Finishing Room.

Day	Concentration C_0 (mg/m^3)				Concentration Estimation Results left (mg/m^3)	
	PM_{10}		$PM_{2,5}$		PM_{10}	$PM_{2,5}$
	Out1	Out2	Out1	Out2		
First	0,61	0,61	1,03	1,64	1,22	2,67
Second	0,67	0,85	1,21	1,82	1,52	3,03
Third	0,79	1,05	1,25	1,74	1,84	2,99
Fourth	0,80	1,00	1,40	1,75	1,80	3,15

Out1 = outdoor 1; Out2 = outdoor 2

Even if the estimation of PM concentration were left behind in the room as though the same every day, so as if there was no PM_{10} and $PM_{2,5}$ were out of the room). However, there was a small portion of PM_{10} and $PM_{2,5}$ concentration that flew out of the room, more or less under $\frac{1}{100}$ mg/m^3. It related to adjusts of the accuracy of the measuring instrument.

3.5. Model Validation with Root Mean Square Error (RMSE)

The root meant square error (RMSE) (Equation 10) was employed to measure the deviation between the random variable with some standard or accepted value. Sivacoumar (2001) examined the error rate of a model was proven to be low if the RMSE value close to zero (Figure 9). The intended random variable was the result of the estimated calculation of PM_{10} and $PM_{2.5}$ concentrations in the room and the accepted value referred to the PM measurement.

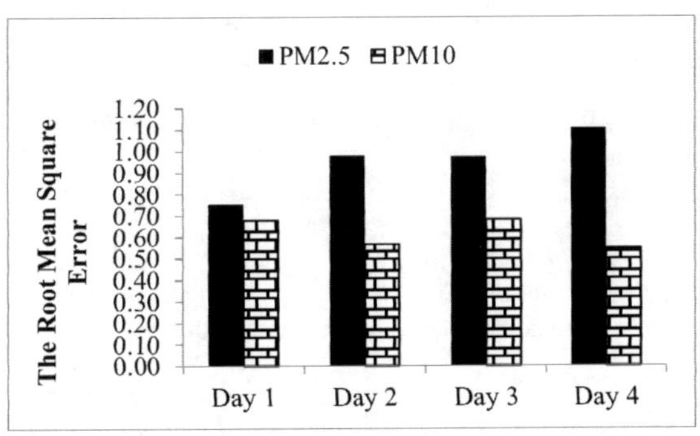

Figure 9. The Root Mean Square Error of PM Concentration.

The RMSE value shows that the bias (error) between PM measurement and the estimated results of PM concentrations was small (close to zero). It meant that the model estimation of PM concentration was close to the actual value (the PM measurement). Giving evidence that the mass balance model was reliable to estimate PM_{10} and $PM_{2,5}$ concentrations.

4. Conclusion

The application of spray-painting operation generates the paint mist that releases to the environment and then flies in the air; thus, it can pollute the environment and endanger human health. Volatile organic compounds (VOCs) from painting solvents are one of the most important sources of pollutant outputs for the shipbuilding and ship repair industry. The concentration of particulates (PM_{10} and $PM_{2,5}$) in the room can be estimated using the mass balance model. The mass balance model is a simple mathematical model that can be used to estimate particulates (PM_{10} and $PM_{2,5}$) in the room. This is evidenced by the value of the root mean square error (RMSE) is small (close to zero), so it can be concluded that the error (error) from the estimated results of particulate concentrations (PM_{10} and $PM_{2,5}$) to the results of small measurements or estimated values close to the actual value (value of measurement results).

5. References

[1] Amir A, Bouzid, Pougnet P and El Hami A 2015 Meta-Model development. *Embedded Mechatronic Systems* **2** pp 151–179 (Elsevier Inc.)
[2] Ugur Bugra C and Vardar N 2008a Investigation of VOC emissions from indoor and outdoor painting processes in shipyards. *Atmospheric Environment* **42** 22 pp 5685–95
[3] Ugur Bugra C and Vardar N 2008b Investigation of VOC emissions from indoor and outdoor painting processes in shipyards. *Atmospheric Environment* **42** 22 pp 5685–95
[4] Cooper C D and Alley F C 2011 *Air Pollution Control: A Design Approach*
[5] Ulrich F, Odeh S, Wiedensohler A, Wehner B and Herbarth O 2011 The effect of particle size on cardiovascular disorders — The smaller the worse. *Science of The Total Environment* **409** 20 pp 4217–21
[6] Zhida H, Ye M, Qingwei N and Ding J 2019 Application of aeration spray technology in China and its reduction effects in VOCs control. *IOP Conf. Series: Materials Science and Engineering* **611** pp 1–8
[7] Inc. Graco 2014 Graco Inc. (Retrieved July 26, 2020) (https://www.graco.com/au/en.html)
[8] Karottki et al 2014 Cardiovascular and lung function in relation to outdoor and indoor exposure to fine and ultrafine particulate matter in middle-aged subjects. *Environment International* **73** pp 372–81
[9] Kelly F J and Fussell J C 2012 Size, source and chemical composition as determinants of toxicity attributable to ambient particulate matter. *Atmospheric Environment* **60** pp 504–26
[10] Lazaridis M, Serfozo N, Chatoutsidou S E and Glytsos T 2015a New particle formation events arising from painting materials in an indoor microenvironment. *Atmospheric Environment* **102** pp 86–95
[11] Lazaridis M, Serfozo N, Chatoutsidou S E and Glytsos T 2015b New particle formation events arising from painting materials in an indoor microenvironment. *Atmospheric Environment* **102** pp 86–95
[12] Myers I. and Maynard R L 2005 Polluted air: Outdoors and indoors. *Occupational Medicine* **55** 6 pp 432–38 (Oxford, England)
[13] NPI 1999a *Emission Estimation Technique Manual for Electroplating and Anodising* (National Pollutant Inventory)
[14] NPI 1999b *Emission Estimation Technique Manual for Surface Coating*
[15] Orona N S, Astort F G A, Maglione P H N, Saldiva J S, Yakisich and Tasat D R 2014 Direct and indirect air particle cytotoxicity in human alveolar epithelial cells. *Toxicology in Vitro* **28** 5 pp 796–802
[16] Santiasih I and Hermana J 2017 A review: The physicochemical characteristics of indoor particulate matters in relation to human health. *ARPN Journal of Engineering and Applied Sciences* **12** 6
[17] Santiasih I, Titah H S and Hermana J 2020 The effects of particulate matters inhalation exposures of prallethrin and D-phenothrin mixture in Mice (Mus Musculus) against exhaled carbon dioxide concentration. *Toxicological Research* **36** 1 pp 59–67
[18] Santiasih I, Titah H S and Hermana J 2019 Carboxylesterase concentration in mouse exposed to particulate matters on inhalation exposure of prallethrin and D-phenothrin mixture. *E3SWC* **125** p 4006
[19] Fredrick S H and Heitner K L 1974 Theoretical model for relating indoor pollutant concentrations to those outside. *Environmental Science and Technology* **8** 5 pp 444–51
[20] SNI 2001 *Tata Cara Perancangan Sistem Ventilasi Dan Pengkondisian Udara Pada Bangunan Gedung*
[21] USEPA 1995 *AP-42: Compilation of Air Emissions Factors, Air Emissions Factors and Quantification.* US EPA (Retrieved June 15, 2020) (https://www.epa.gov/air-emissions-factors-and-quantification/ap-42-compilation-air-emissions-factors)
[22] USEPA 1997 *Environmental Protection Agency 40 CFR Part 50 National Ambient Air Quality Standards for Particulate Matter; Final Rule*
[23] Wadden R A and Scheff P A 1983 *Indoor Air Pollution: Characterization, Prediction, and Control.* (Wiley)

IOP Conf. Series: Materials Science and Engineering 1175 (2021) 012004 doi:10.1088/1757-899X/1175/1/012004

Design of android based Unmanned Surface Vehicles (USV) for oil spill monitoring

Ardiansyah[1], I Z Oktaviano[2], I Sutrisno[3], R Y Adhitya[4], A Z Arfianto[5], D D Suharso[6], I Suwondo[7], V Y P Ardhana[8], H Soeroso[9] and M D Mulyodiputro[10]
[1,6,7]Politeknik Pelayaran Surabaya, Indonesia
[2,3,4,5,9]Politeknik Perkapalan Negeri Surabaya, Indonesia
[8,10]Universitas Qamarul Huda Badaruddin, Indonesia

E-mail: ardi9415096@gmail.com

Abstract. In this research, a USV will be developed that can detect oil spills on the waters, so in the future it can overcome the problem of water pollution. To be able to determine the thickness of the oil on the water surface, TEMT6000 and LED sensors are used. The light intensity will be converted into ADC value by the TEMT6000 sensor. Furthermore, on this system USV will be controlled by Android and can monitor water surface conditions. Android applications will be connected to Wemos via wifi. Then Wemos is connected serially with a RF915 and another RF915 will be connected to the microcontroller of the USV. From the results of the USV test, the USV can be controlled via Android by moving forward, backward, turning left and right. The maximum range of RF915 is 493.3 meters. Then the oil spill detection average error is 5.45% and the the average of the success rate is 94.55%.

1. Introduction

Recently the industry is experiencing a very rapid development. With the increasing number of industrial companies, the more waste is produced which has the potential to pollute the environment. Industrial waste is waste originating from industrial activity discharges which can be in the form of solid, liquid, or gas. If the waste is not treated first, it can pollute the environment which can have a negative impact on the ecosystem of living things around the waste disposal site.

Apart from industrial waste, environmental pollution can also be caused by work accidents resulting in pipe leaks, sinking of oil-carrying vessels which results in oil spilling. According to data from the Ministry of Energy and Mineral Resources from 2013 to 2017, the highest oil spill was in 2017 with 2071 barrels and the lowest in 2014 with 46 barrels [1]. The number of oil spills increased throughout 2018 and 2019. On January 10, 2019, a tanker owned by PT Sochi experienced an oil leak that polluted the waters in Pare-Pare, South Sulawesi [2]. This can be detrimental to local fishermen who cannot go to ocean because of the oil spill in their fishing area.

In the same year, PT PERTAMINA's pipeline leak resulted in an oil spill of 39 thousand barrels and 5 million sacks in Karawang [3]. In the short term, an oil spill in Karawang can cause a decrease in the quality of ocean and coastal water, the death of several types of fish and shellfish [4]. The problem of oil spills must be addressed before it causes the death of the marine biota. Apart from being at ocean, a leaky pipe belonging to PT PERTAMINA in the Tasikmalaya area also caused 45 residents' fishponds to be polluted by diesel fuel [5]. This also gives quite a lot of losses for people who have the main job as a owners of fishing ponds and fish breeders.

Content from this work may be used under the terms of the Creative Commons Attribution 3.0 licence. Any further distribution of this work must maintain attribution to the author(s) and the title of the work, journal citation and DOI.
Published under licence by IOP Publishing Ltd

For this reason, a right technique is needed to detect an oil spill. In previous research, a USV (Unmanned Surface Vehicle) was made that can monitor water quality with pH as a parameter [6]. In this paper, a design and development of a oil spill detection using an Android-based USV (Unmanned Surface Vehicle) is expected to solve the problem of oil spills in the water if there is further development in the future.

In 2013, research conducted by Sangwoo Oh and Moonjin Lee entitled Detection of Hydrocarbon Oil in Seawater by Light Absorption Analysis. In this research, a sensor design that can detect the presence of oil spills and its thickness on the water surface is made using a photodiode and LED sensor [7]. If there is oil on the surface of the water, the light intensity emitted by the LED also decreases. Then the thickness of the oil can be predicted through an LDR sensor that detects the light intensity from the LED.

In 2018, research conducted by Bagus Aris Saputra et al entitled Design and Build of Water-Oil Separation Systems Based on Microcontroller Teensy. In this study the design was made to detect the presence of oil on the water can be done by adding a laser from below the surface of the water on the ldr sensor which is above the surface of the water, then carried out oil absorption using a diaphragm pump. However, this system has not been applied to the USV (Unmanned Surface Vehicle) which can move to absorb oil, the thickness of the oil cannot be monitored using a laptop or smartphone [8].

In 2019, research conducted by Jo et al with the title A low-cost and small USV platform for water quality monitoring. In this research, a design of USV (Unmanned Surface Vehicle) was made which is used to monitor water quality with parameters of pH, temperature and turbidity. This system can be used to make it easier for researchers to monitor river water conditions around industrial areas [6].

Based on previous research, the authors developed a USV prototype to detect spill. This study develops the first research, then applies it to USV and the detected thickness data can be displayed on the android application.

2. Methodology

2.1.System Design

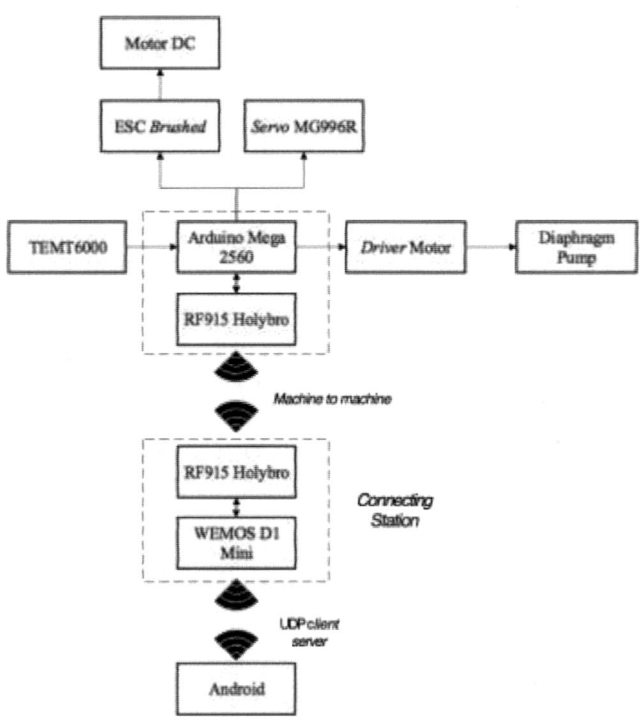

Figure 1. System Design.

Figure 1 shows the system design. To send data from Android to USV, Android sends data to Wemos from via wifi with the UDP protocol. Furthermore, Wemos will send data obtained from Android to RF915 via serial communication. Then RF915 will send data to RF915 which is located on the USV and the data is read by a microcontroller with serial communication. To send sensor data applies the opposite, starting from the microcontroller reading the TEMT6000 sensor and sending data serially to the RF915. Then the RF915 from the ship will send the data to the connecting station RF915. Then the RF915 will send the data serially to Wemos. Finally Wemos sends data to Android via wifi with UDP communication.

2.2. USV Design

The USV (Unmanned Surface Vehicle) that will be made has 3 main dimensions, 70 cm long, 35 cm wide and 11 cm high.

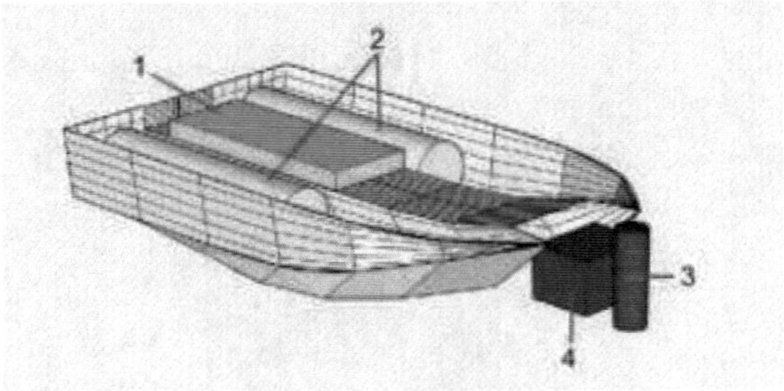

Figure 2. Isometric View of USV Hardware Design.

From some of the images above, Figure 2 shows the initial USV design view isometric. In accordance with the image above, the following is an explanation of the placement of components on the USV:

1. The electric circuit is a circuit consisting of a microcontroller and an actuator which functions to move the ship, read the sensor and carry out oil extraction.
2. The tank will be placed on both hulls of the ship to store oil that has been sucked from the water surface.
3. The sensor on the USV is a design of the TEMT6000 sensor that detects the light intensity of the LED that shoots light at the TEMT6000. The position of the LED is below the water level and the TEMT6000 is above the water level.
4. The buoy functions to support the hose and the TEMT 6000 sensor circuit which will suck up oil spill samples.

2.3. Sensor Placement Design

To measure oil thickness using LEDs and TEMT6000, the laser is placed in the water and panned to the surface of the water. The laser light intensity will be measured by the TEMT6000. The oil thickness above the water surface will affect the light intensity received by TEMT6000, then this is used to predict the thickness of the oil. Figure 3 below is the layout design for the LED and TEMT6000.

IOP Conf. Series: Materials Science and Engineering 1175 (2021) 012004 doi:10.1088/1757-899X/1175/1/012004

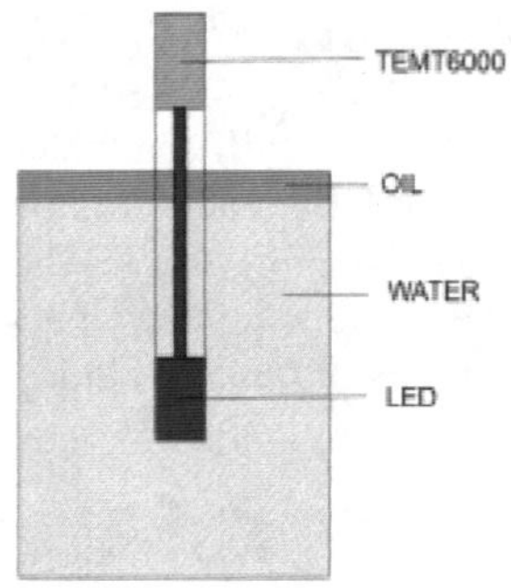

Figure 3. Placement of the LED and TEMT6000.

2.4.Android App Design

The android application that will be used has 2 main pages, namely monitoring and control as shown in Figure 6 and Figure 7. Then when you open it for the first time, a loading page will appear as shown in Figure 4. To switch from the control page to monitoring as well as preferably there is a menu page as shown in Figure 5.

Figure 4. Android Application Loading Page.

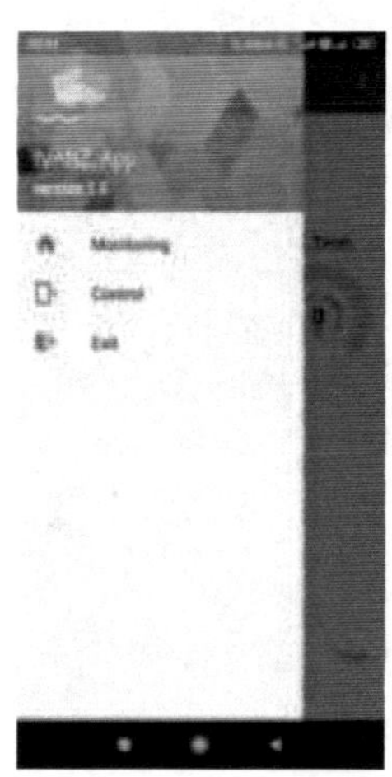

Figure 5. Android Application Menu Page.

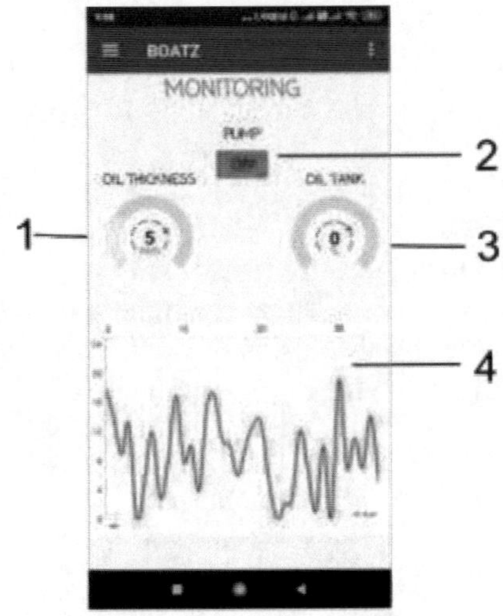

Figure 6. Android Application Monitoring Page.

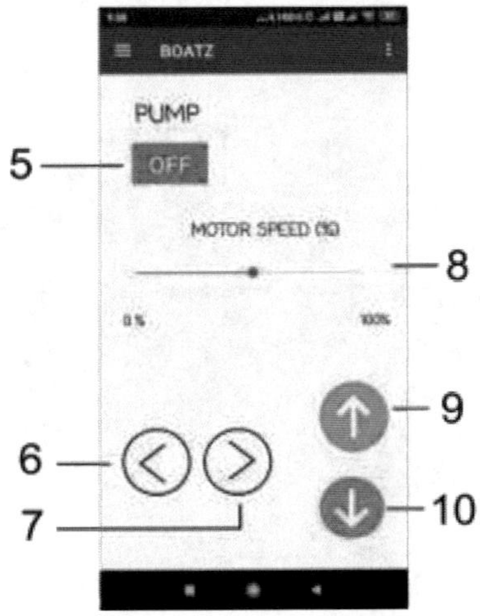

Figure 7. Android Application Control Pages.

Following is the description according to the interface design in Figure 9 and Figure 10:
1. A gauge that displays the detected oil thickness in millimeters (mm).
2. The oil pump indicator shows the condition of the oil suction pump on or off.
3. Indicator of tankage on vessels with a range of 0 100%.
4. Graph of the detected oil thickness in millimeters (mm).
5. Button to activate and deactivate the oil-suction pump.
6. Button to move the ship to turn left.
7. Button to move the ship moving to the right.
8. Seekbar to set the boat speed from 0% to 100%.
9. Button to move the ship forward.
10. Button to move the ship backwards.

3. Result and Discussion

3.1. USV Design Result

USV is the main hardware in this final project, this USV has a catamaran type or two right and left hulls. This USV is made of balsa wood. The specifications of this USV include the following:

Length	: 75 cm
Width	: 40 cm
Height	: 11 cm
Oil Tank Capacity	: 2 Liters

The reason for using the catamaran hull is that the right hull and left hull can be installed with an oil transport tank to take oil samples by absorpsion method. This is shown on the inside of the USV as Figure 8. Following Figure 9 which shows the isometric view of the USV.

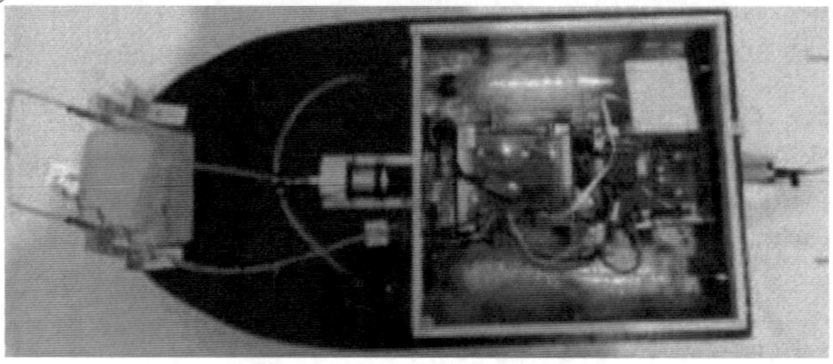

Figure 8. Inside the USV.

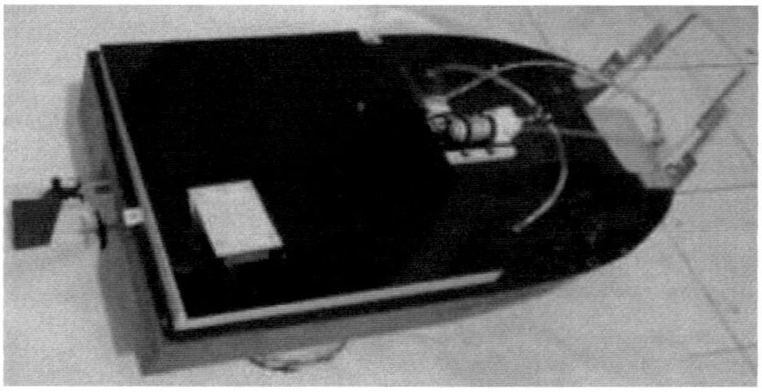

Figure 9. USV Isometric View.

As shown in Figure 8 and Figure 9, the front of the USV has a buoy which functions to support the TEMT6000 sensor circuit to float on the water surface so that it can detect oil spills on the water surface. In addition, the float also functions to support the suction hose so that it can suck up oil samples. The buoy can move up and down so that when the water laden from USV rises due to oil holding, the position of the TEMT6000 sensor circuit and the oil suction hose remain in the same position on the water surface.

3.2. RF915 Holybro Range Test Result

This test is carried out by connecting a Holybro RF915 with a powerbank using a USB cable while the other Holybro RF915 is already in the connecting station as in Figure 10. Then wait until the two are connected to each other with the green LED indicator lights up constantly. The connecting station is placed in the fixed place then the coordinates are recorded with GPS, then the RF915 Holybro which is connected to the powerbank is taken to another place by vehicle until the two radios are not con-

nected. Furthermore, the GPS coordinates where the last time the two radios were still connected were recorded. Here are the coordinates of the two radios and the distance is about 493.30 meters as in Figure 11.
1st Holybro RF915 = -7.2700309,112.8048667
2nd Holybro RF915 = -7.2657331,112.8037614

Figure 10. Connecting Station and RF915 Test.

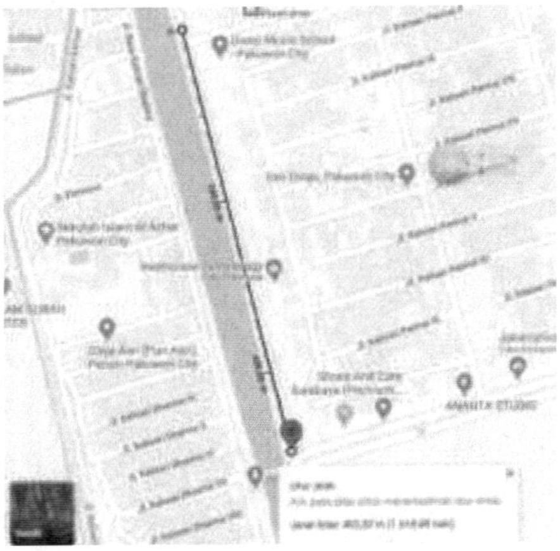

Figure 11. Connecting Station and RF915 Test Result.

3.3. Oil Spill Detection Result

This test is done by spilling oil (diesel fuel) on the water in the aquarium. Oil will be spilled on the surface of the water with a certain thickness then ADC data will be recorded. Figure 12 shows the TEMT6000 sensor testing to retrieve ADC data from the light intensity emitted by the LED. The following Table 1 is the result of ADC data from the taking made on the TEMT6000 sensor.

IOP Conf. Series: Materials Science and Engineering 1175 (2021) 012004 doi:10.1088/1757-899X/1175/1/012004

Figure 12. TEMT6000 Sensor Test.

Table 1. Results of TEMT6000 Sensor Data Retrieval.

ADC	Ruler(mm)
83	20
102	18
126	16
141	14
157	12
188	10
202	8
224	6
236	4
251	2
278	0

As in the experimental results above, the ADC value of the TEMT 6000 sensor is directly proportional to the intensity of the light hitting the sensor. The higher the light intensity, the greater the ADC value.

The graph in Figure 13 shows the relation between oil thickness and ADC that is read quite linear. After obtaining this data, the data is made a graph of the response to the effect of oil thickness on ADC which is read by the TEMT6000 sensor as in Figure 13 where the x-axis is the ADC value and the y-axis is the thickness of the oil. From this graph, using Microsoft Excel, we get a trendline with functions such as Equation 1 below:

$$y = -0.1037x + 28{,}748 \quad (1)$$

IOP Conf. Series: Materials Science and Engineering 1175 (2021) 012004 doi:10.1088/1757-899X/1175/1/012004

With:
y = oil thickness (mm)
x = TEMT6000 ADC output

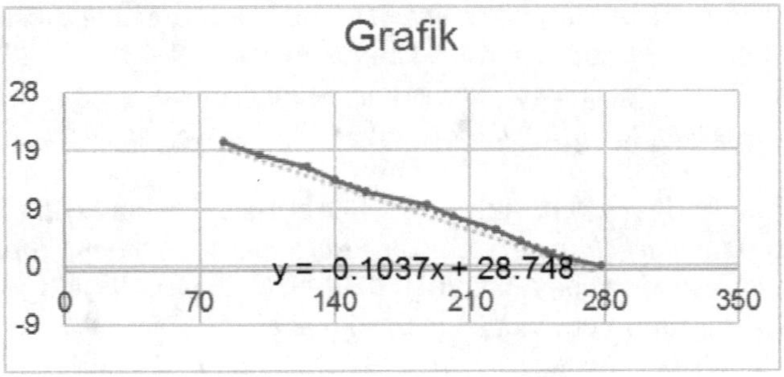

Figure 13. Oil Thickness vs ADC Value Graph.

Then the ADC value is converted to oil thickness according to Equation 1. The following Table 2 is a table of detection results with the errors.

Table 2. Result of Oil Spill Detection.

Ruler (mm)	Sensor (mm)	Error (%)
20	20	0
18	18	0
16	16	0
14	14	0
12	12	0
10	9	10
8	8	0
6	6	0
4	4	0
2	3	50
0	0	0
Average Error (%)		5.45

4. Conclusion

The conclusion is that USV can be controlled via Android by moving forward, backward, turn left and turn right. Maximum range of RF915 is 493.3 meters. Then the oil detection system average error is 5.45% and the average of success rate is 94.55%.

5. References

[1] Kementrian ESDM 2017 *Data Tumpahan Minyak*. Available: http://statistik.migas.esdm.go.id/index.php?r=dataTumpahanMinyak/index (Accessed 26-Jan-2020)

[2] Suara.com 2019 *Pertamina dan PT Soechi Didesak Bertanggungjawab Kasus Kapal Tanker Bocor*. Available: https://www.suara.com/news/2019/01/25/223344/pertamina-dan-pt-soechi-didesak-bertanggungjawab-kasus-kapal-tanker-bocor (Accessed 09-Jan-2020)

[3] Arvirianty A 2019 *Tumpahan Minyak Pertamina Capai 39 Ribu Barel & 5 Juta Karung*. Available: https://www.cnbcindonesia.com/news/20190920133342-4-100979/tumpahan-minyak-pertamina-capai-39-ribu-barel-5-juta-karung (Accessed 26-Jan-2019)

[4] CNN 2019 *Kandungan Minyak Bocor Pertamina Ancam Biota Laut*. Available: https://www.cnnindonesia.com/teknologi/20190816123819-199-421892/kandungan-minyak-bocor-pertamina-ancam-biota-laut (Accessed 09-Jan-2020)

[5] Tribunnews.com 2014 *Sumur dan Kolam Ikan Tercemar Solar*. Available: https://www.tribunnews.com/regional/2014/10/23/sumur-dan-kolam-ikan-tercemar-solar (Accessed 09-Jan-2020)

[6] Jo W, Hoashi Y, Paredes Aguilar L L, Postigo-Malaga M, Garcia-Bravo J M and Min B C 2019 A low-cost and small USV platform for water quality monitoring. *HardwareX* **6** p. e00076

[7] Oh S, Lee M and Principle A S 2013 Detection of hydrocarbon oil in seawater by light absorption analysis. *SENSORDEVICES Fourth Int. Conf. Sens. Device Technol. Appl. Detect* No. c pp 103–106

[8] Saputra B A, Rivai M and Tasripan T 2019 Rancang bangun sistem pemisah air – Minyak berbasis metode adsorpsi menggunakan mikrokontroler Teensy. *J.Tek. ITS* **7** 2

[9] Jami'in M A, Hu J, Marhaban M H, Sutrisno I and Mariun N B 2016 Quasi ARX neural network based adaptive predictive control for nonlinear systems. *IEEJ Trans Elec Electron Eng* **11** pp 83-90

[10] Jami'in M A, Sutrisno I and Hu J 2012 Lyapunov learning algorithm for Quasi-ARX neural network to identification of nonlinear dynamical system. *IEEE International Conference on Systems, Man, and Cybernetics (SMC)* (Seoul, 2012) pp 3147-3152 doi: 10.1109/ICSMC.2012.6378275

[11] Jami'in M A, Sutrisno I and Hu J 2013 Deep searching for parameter estimation of the linear time invariant (LTI) system by using Quasi-ARX neural network. *The 2013 International Joint Conference on Neural Networks (IJCNN)* (Dallas-TX, 2013) pp 1-5 doi: 10.1109/IJCNN.2013.6707091

[12] Jami'in M A, Sutrisno I and Hu J 2015 Maximum power tracking control for a wind energy conversion system based on a quasi ARX neural network model. *IEEJ Trans Elec Electron Eng* **10** pp 368-375

[13] Jami'in M A, Sutrisno I, Hu J, Mariun N B and Marhaban M H 2014 An adaptive predictive control based on a quasi-ARX neural network model. *13th International Conference on Control Automation Robotics & Vision (ICARCV)* (Singapore, 2014) pp 253-258 doi: 10.1109/ICARCV.2014.7064314

[14] Sutrisno I, Jami'in M A and Hu J 2014 Modified fuzzy adaptive controller applied to nonlinear systems modeled under quasi-ARX neural network. *Artif Life Robotics* **19** pp 22–26 https://doi.org/10.1007/s10015-013-0137-6

[15] Sutrisno I, Jami'in M A and Hu J 2014 An improved elman neural network controller based on quasi ARX neural network for nonlinear systems. *IEEJ Trans Elec Electron Eng* **9** pp 494-501

[16] Sutrisno I, Che C and Hu J 2014 An improved adaptive switching control based on quasi-ARX neural network for nonlinear systems. *Artif Life Robotics* **19** pp 347–353

[17] Sutrisno I, Jami'in M A and Hu J 2012 Neural predictive controller of nonlinear systems based on quasi-ARX neural network. *18th International Conference on Automation and Compu-*

ting (ICAC) (Loughborough, 2012) pp 1-6

[18] Widodo R B et al. Grasping and attached mode in human computer interaction in the study of mouse substitution. *Journal of Physics: Conference Series* **1196** 1 012034

[19] Sutrisno I, Jami'in M A, Hu J and Marhaban M H 2015 Self-organizing quasi-linear ARX RBFN modeling for identification and control of nonlinear systems. *54th annual conference of the society of instrument and control engineers of Japan (SICE)*. IEEE

[20] Jami'in M A, Sutrisno I and Hu J 2014 Nonlinear adaptive control for windenergy conversion systems based on quasi-ARX neural networksmodel. *Procedings of the International Multi-Conference of Engineers and Computer Scientists* (March 12-14, 2014) (Hong Kong)

[21] Sutrisno I 2009 *Pemrograman Komputer dengan Software Matlab* (Surabaya: ITS Press)

IOP Conf. Series: Materials Science and Engineering 1175 (2021) 012005 doi:10.1088/1757-899X/1175/1/012005

Dynamic time distribution system monitoring on traffic light using image processing and convolutional neural network method

M Rifai[1], R A Budiman[2], I Sutrisno[3], A Khumaidi[4], V Y P Ardhana[5], H Rosika[6], Tibyani[7], M Setiyono[8], F Muhammad[9], M Rusmin[10] and A Fahrizal[11]
[1]Politeknik Penerbangan Surabaya, Indonesia
[2,3,4]Politeknik Perkapalan Negeri Surabaya, Indonesia
[5,6]Universitas Qamarul Huda Badaruddin, Indonesia
[7]Brawijaya University
[8,9]Politeknik Pelayaran Sorong, Indonesia
[10,11]Universitas Muhammadiyah Sorong, Indonesia

E-mail: rifai.moch.smart@gmail.com

Abstract. The number of vehicles that continues to increase every time can cause congestion and increase the potential for accidents. One type of violation that often results in accidents is motorists who break through traffic lights when the red light is on. One factor that encourage motorists to not obey traffic light signs is due to the distribution of green time which tends to be fast on relatively dense tracks. This final project aims to get dynamic green light on traffic light according to the volume of the vehicle for each lane using convolutional neural network method and can detect traffic light violations in the form of breaking the stop line markings on zebra cross. The sensor used is camera. There are three cameras used in each line. The number of cars detected will be sent to microcontroller via laptop. The data on the number of cars is used as input for the convolutional neural network to produce output data that serves as reference for obtaining the length of time for the green light. If the car stops above or exceeds the stop line on zebra cross when the red light is on, the car will be detected as violation.

1. Introduction

One of the effects of the increasing population is that the number of vehicles on the road is also increasing. Based on data from the Central Statistics Agency in 2015, 2016, 2017 in a row the number of motor vehicles was 121,394,185, 129,281,079, 138,556,699 vehicles. This is not proportional to the volume of roads which tend to be fixed, this results in traffic congestion which results in congestion. Congestion can interfere with the mobility of the community.

The amount of time lost in wasted due to congestion has an impact on people's income. Congestion cannot be avoided when the hours go to work or go to school. That also applies when the hours come home from work or school. Congestion is also affected by sub-optimal traffic light settings and traffic violations by motorists. Traffic violations are caused by motorists who do not obey traffic signs. One of the traffic violations is that the driver stops on the zebra crossing on the traffic light which is used by pedestrians to cross the road. The Bandung City Transportation Agency released a type of violation that was often violated by motorists in the period 15 to 27 November 2017, stopping at the zebra

Content from this work may be used under the terms of the Creative Commons Attribution 3.0 licence. Any further distribution of this work must maintain attribution to the author(s) and the title of the work, journal citation and DOI.
Published under licence by IOP Publishing Ltd

IOP Conf. Series: Materials Science and Engineering 1175 (2021) 012005 doi:10.1088/1757-899X/1175/1/012005

crossing on the traffic light was the most frequently committed violation of 326 violations (Mawardi, 2017).

The ideal condition (time) of traffic lights distribution is important which can reduce the density of vehicles by adjusting the distribution which can adjust according to the level of congestion at each time. So it is necessary to have a system of tools that can detect violations and can adjust the state (time) of traffic lights according to the density level of the vehicle. Research on getting dynamic time has been done several times. The research that has been done is in the form of analysis and design simulation using a microcontroller and PLC. Based on the above background, the authors took the research to create a dynamic time distribution system that is suitable for congested traffic conditions and detects violations when the vehicle's position crosses the crossing markings when the traffic light is red.

The author makes a design using the Convolutional Neural Network (CNN) based image processing method, so that the results of the research conducted can be known. Therefore, the authors raised a research entitled " Dynamic Time Distribution System Monitoring on Traffic Light Using Image Processing and Convolutional Neural Network Method".

2. Relevant Study

2.1. Related Work

The research entitled "Design of Dynamic Time Distribution System for Traffic Light Using Image Processing Method Based on Fuzzy Logic" has studied calculating traffic density using camera sensors and can control the time at traffic lights. In this study, we cannot distinguish between left and right paths. The counted vehicle is the total number of vehicles on the road. (Hidayat, 2018)

Subsequent research entitled "Design and Development of Dynamic Time Distribution Systems and Detection of Violations in Traffic Light Using Fuzzy Logic Based Image Processing Methods has conducted research on calculating the density of traffic lanes and detection of zebracross markers violations using camera sensors and can control time at traffic lights. In this study, it cannot detect objects other than cars. Vehicle detection is based solely on color and cannot detect smaller objects. (Saputro, 2019)

In this study, referring to previous research developed where researchers used the Image Processing method and the Convolutional Neural Network method. In contrast to previous studies, in this study the method used was Convolutional Neural Network. This convolutional neural network method is able to detect objects based on the entered datasheet. Image Processing is used to determine the number of vehicles on the left lane (Hidayat, 2018).

3. Research Design

3.1. System Design

This chapter contains the research flow. For the discussion that is about the identification of problems, study of literature, system design, hardware and software design, analysis and discussion. Systematically the steps in the research are made in the form of flowcharts as shown in Figure 1 below.

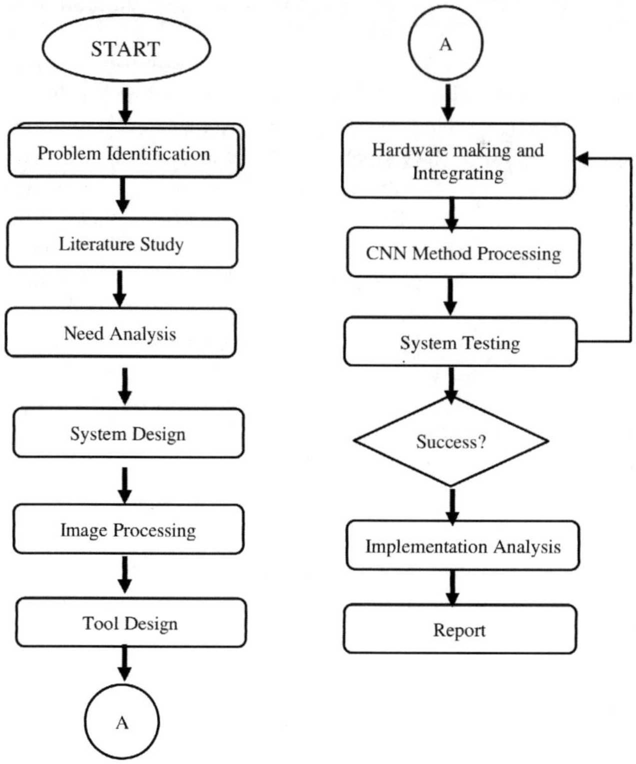

Figure 1. System Design.

3.2. Design

Dataset is an object that represents data and relations that have been created in storage. A dataset can be called a database, because it has the same structure. The dataset in this study is taken from several examples of several images that have been stored in the laptop storage space. And the dataset is used for object classification. the more input the dataset will make the higher the classification results and the heavier the laptop runs.

As in Figure 2 above, some examples of dataset images and the number of datasets used. However, later the dataset used will increase even more.

Dataset Name	Dataset Image	Total
Stuck		200
Middle		200
Slack		200

Figure 2. Dataset.

Hardware placement designs such as PC placement and camera sensors are designed as best as possible in order to get an accurate and efficient system response. As planned as shown in Figure 3 and 4 below.

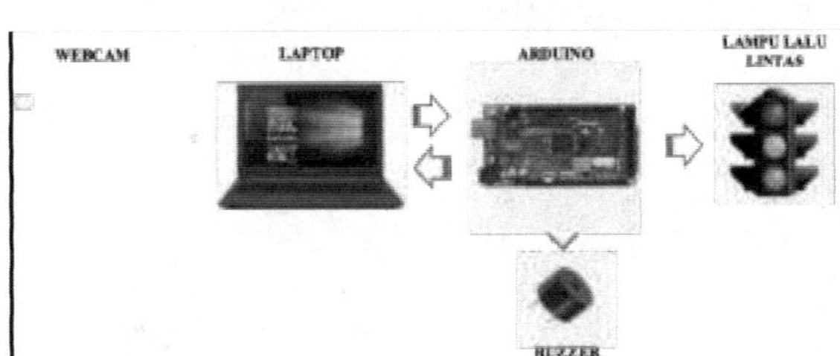

Figure 3. Hardware Design System.

Figure 4. Tool Design.

3.3.CNN Architecture Design

This final project is focused on getting an object detection system with a Mini PC. In addition, the main objective of this study is to obtain a classification result from the CNN process. In the general CNN process, this process has 3 stages, namely pre-processing, processing, and classifying.

The pre-processing process consists of 2 processes, namely the formation of inputs and conversion of inputs to grayscale. The second process is processing which consists of image convolution, image dimension reduction, max pooling, training with neural networks, and softmax. For the last process is classifying which has 1 process, namely determining output.

As in Figure 5 with the flow of the feature extraction process, namely by using the convolution process which is carried out 2 times, with the subsampling layer only done once. And for the stages, the first stage is the grayscale image which is processed through a convolution kernel or can be called a 5 x 5 Gaussian kernel with 10 convolution features to produce 10 new images of the same size without reducing pixels, 320 x 240 pixels, then the convolution process is repeated for the The second time with a 5 x 5 kernel in this convolution the feature map does not increase because each feature map is only convolved once so that the second feature map is 316 x 236 x 10. The next process is subsampling and ReLu with half reduction. The pixel size of the previous image, therefore, on this layer the feature map will be 158 x 118 x 10. After the feature extraction process is done, flatting is carried out, namely making the 2 D (mxn) matrix become 1D (mx 1), this flatting process produces an input vector of 186,440 who will then go through the training process with several similar output vectors the shape. For the image in 3.10 with a car input and there are 4 output vectors that resemble, namely truck, car, tv monitor, chair, motorcycle. The training is carried out with the help of softmax. And the output turned out to be the correct one is car.

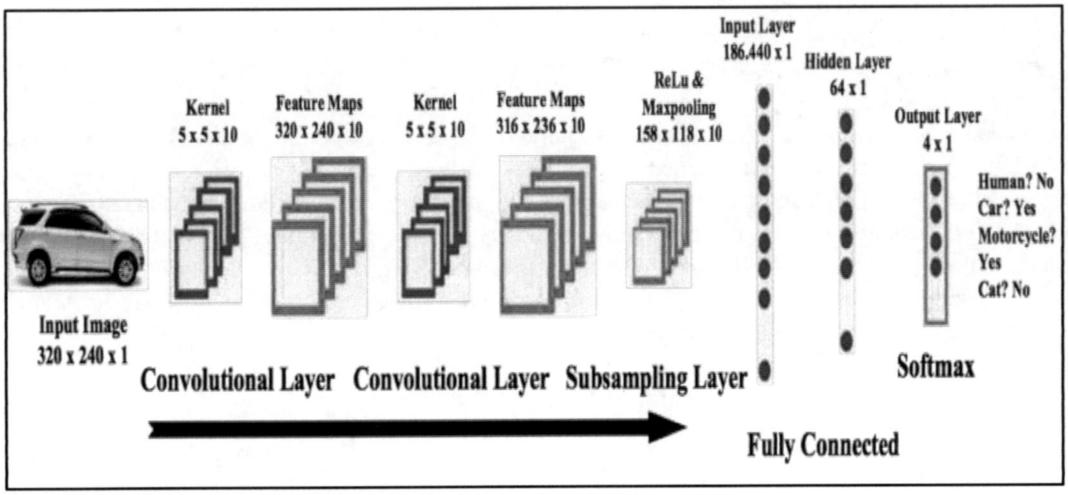

Figure 5. CNN Process Architecture Design.

3.4. Image Processing Design

In image processing, there is an initial process called pre-processing which is used to obtain information about the original image that is ready to be processed for the next stage, which can be called processing so that the result is obtained.

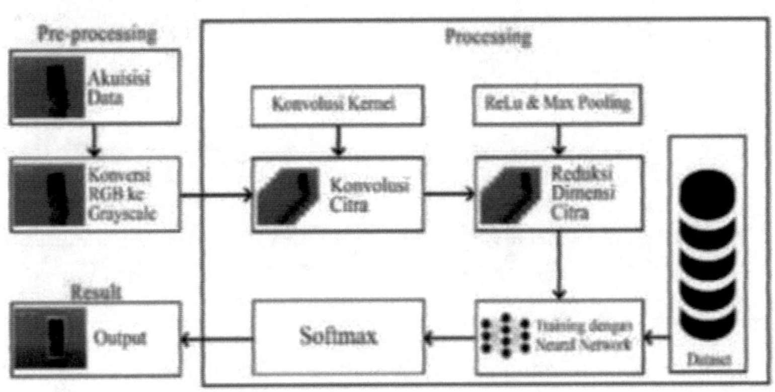

Figure 6. CNN Process Sketch.

Figure 6 is a general sketch of the process of the CNN method. Before entering the CNN stages, the quantity and quality of the dataset must be determined first. Because it will affect the results of the training. The pre-processing stage is the process of obtaining image data which is then converted from an RGB image to Grayscale. Whereas at the Processing stage, an image convolution process occurs which functions to improve the image by means of a filter using the kernel convolution so that the im-

IOP Conf. Series: Materials Science and Engineering 1175 (2021) 012005 doi:10.1088/1757-899X/1175/1/012005

age looks more detailed by reducing noise. Then proceed with dimension reduction to change the image pixel size from the convolution process in half. After obtaining an image of the next reduction process, the result of the reduction is trained by the neural network to obtain the weight. This weight can determine and produce the Output value on the system.

4. Result and Discussion

4.1. Sensor and Actuator Testing

4.1.1. Camera Sensor Testing.

To proceed to the next stage, a sensor test is required to be installed in this system. The following are the results of sensor testing in this Final Project plan. In this section, contains about camera testing. For checking the camera can be seen directly from the indicators contained on the camera. There is a green LED indicator located in the front area. This indicator will light up when the program asks the camera to turn on. In the application in the system, the camera has a little time to start up at the beginning. In 1, the following is an example of an indicator when the camera is on and off. And in Table 1 the following is the time it takes for the camera to turn on.

Table 1. Webcam Condition.

No	Webcam On	Webcam Off
1	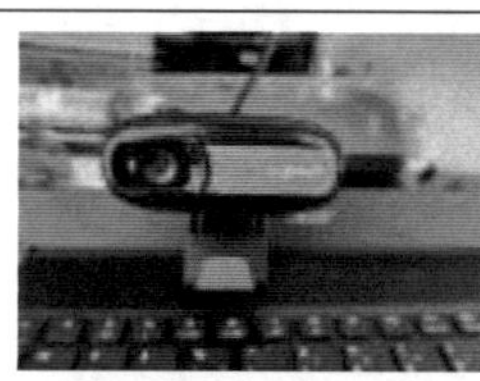	
2		
3		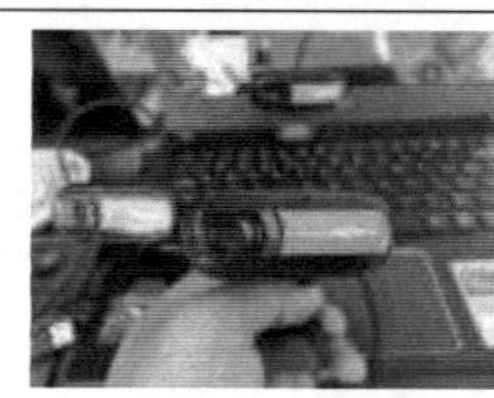

The next step is testing the webcam connection to the PC. The PC will be given a program that will read and display the video in real-time. If successful, the test of the camera connection is complete. Figure 7 shows the results of testing the connection between the camera and PC.

IOP Conf. Series: Materials Science and Engineering 1175 (2021) 012005 doi:10.1088/1757-899X/1175/1/012005

Figure 7. Testing Webcam to PC.

4.1.2.Pilot Lamp Testing.

To proceed to the next stage, it is necessary to test the actuatator that will be installed in this system. The following are the results of actuator testing in this Final Project plan. In this section, contains about testing the Pilot Lamp. To check the Pilot Lamp, it can be seen directly from the flame of the lamp. The following is an example of an indicator when the Pilot Lamp turns on and off. In Table 2 the following is an image when the Pilot Lamp is On, and an image when the pilot lamp turns off.

Table 2. Pilot lamp Condition.

No	Pilot Lamp On	Pilot Lamp Off
1		
2	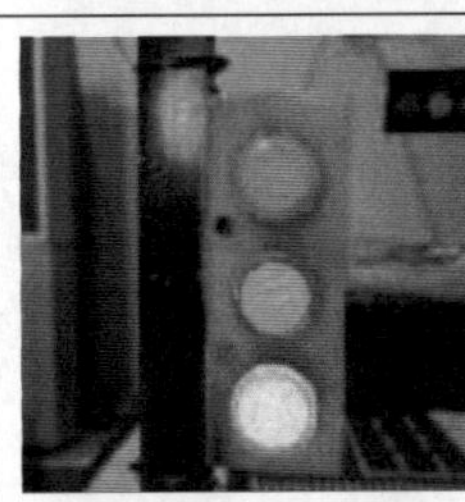	
3		

4.2. *Training Result*

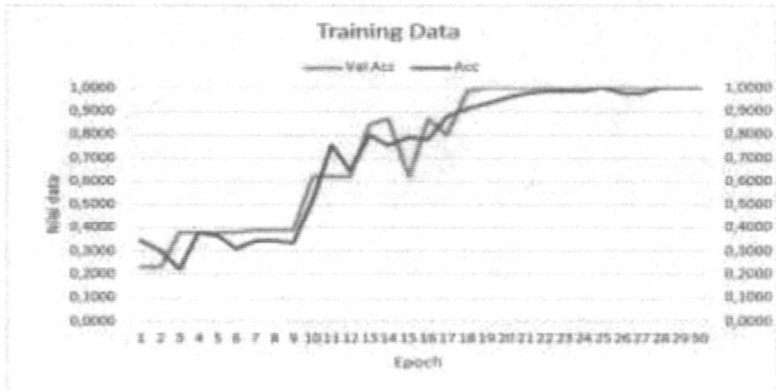

Figure 8. Dataset Training Result.

Figure 8 is the result of training using the aldadelta optimizer. From the graph, it shows that the results obtained using the aldadelta optimizer have an accuracy of up to 1 using 30 epochs.

4.3. *Result Image Processing and CNN*

Table 3. Test Result.

No	Figure Detected	Monitoring
1		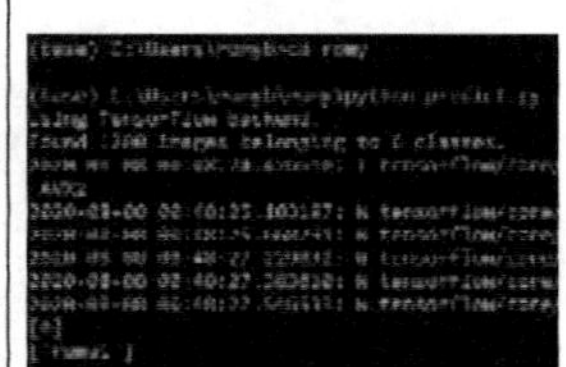
2		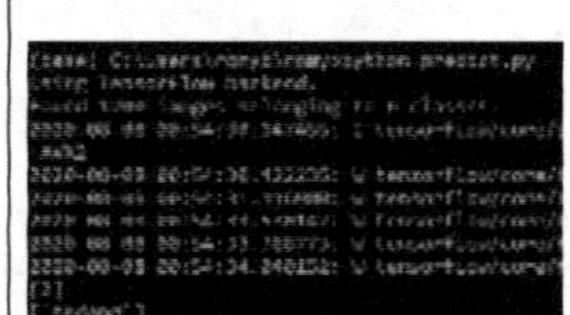
3	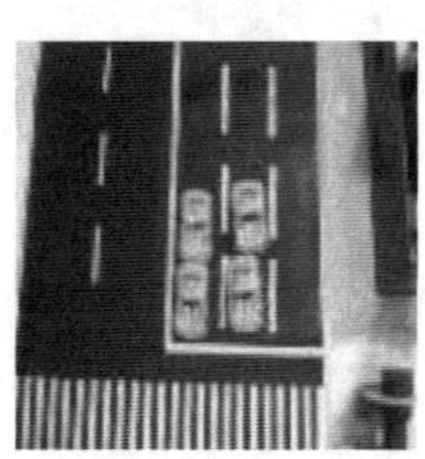	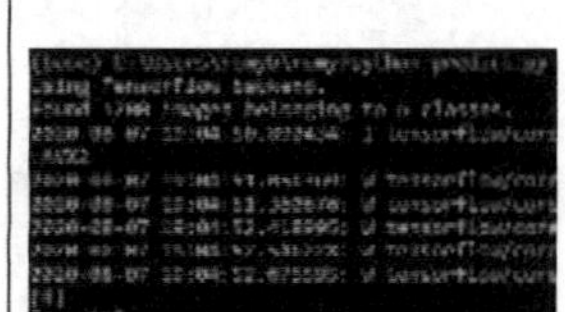

4.4. Hardware

The mechanical design in this tool is made using wood as a base, while the road image uses a banner with a size of 110 cm x 110 cm. The intersection used is a 4-lane intersection, one of which is a directional route. There are 3 traffic light poles and there is 1 camera on each traffic pole that is used to detect the number of vehicles in 3 traffic lanes. Figure 9 is the prototype used in this final project. The prototype is based on the mechanical design in chapter 3.

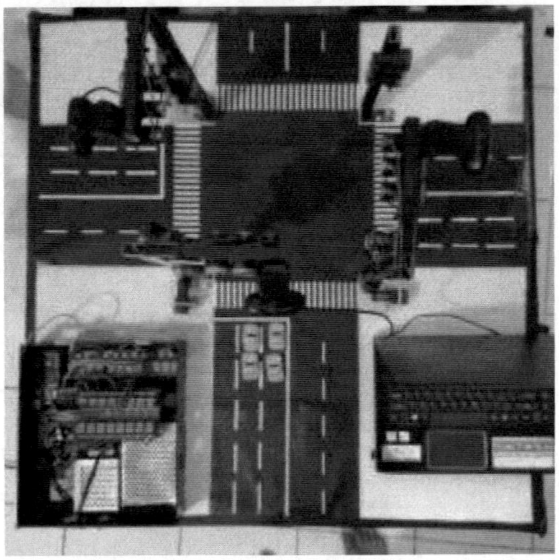

Figure 9. Hardware system.

5. Conclusion

The conclusion of this study that this Convolutional Neural Network method can monitor density levels well. The existing dataset already has a good level of accuracy, reaching 100%. This level of accuracy can be seen from the results of training data that have been carried out using the Adadelta optimizer.

6. References

[1] Khumaidi A 2017 Welding defect classification based on Convolution Neural Network (CNN) and Gaussian Kernal. *International Seminar on Intelligent Technology and Its Application* pp 261-265

[2] Deshpande A 2016 *A Beginner's Guide to Understanding Convolutional Neural Network Part 2*

[3] Suartika I W E 2016 Klasifikasi citra menggunakan Convolutional Neural Network (CNN) pada Caltech 101. *Jurnal Teknik ITS* **5** A65-A69

[4] Limantoro S, Kristian Y and Purwanto D 2017 *Deteksi Pengendara Sepeda Motor Menggunakan Deep Convolutional Neural Networks* (Teknologi Informasi Sekolah Tinggi Teknik Surabaya)

[5] Lutfi M 2018 *Pendeteksi Objek Api dan Boneka pada Robot Pemadam Api Berkaki Menggunakan Video Processing Berbasis Mini PC* (Politeknik Perkapalan Negeri Surabaya)

[6] Rappler.idntimes.com 2018 *This Able Enterprise Memberdayakan Penyandang Disabilitas*. (Available: https://rappler.idntimes.com/sakinah-haniy/thisable-enterprise-memberdayakan-penyandang-disabilitas)
[7] Sena S 2017 *Pengenalan Deep Learning Part 7: Convolutional Neural Network*. (Available: https://medium.com/@samuelsena/pengenalan-deep-learning-part-7-convolutional-neural-network-cnn-b003b477dc94)
[8] Sidharta H 2017 *Introduction to Open* (CV Ilmu Komunikasi Binus Malang)
[9] Soebhakti H 2009 TPA81 Thermopile Array. *AVR Application Note* pp 1-6
[10] Stanford University 2016 *An Introduction to Convolutional Neural* (University Stanford).
[11] Jami'in M A, Hu J, Marhaban M H, Sutrisno I and Mariun N B 2016 Quasi ARX neural network based adaptive predictive control for nonlinear systems. *IEEJ Trans Elec Electron Eng* **11** pp 83-90
[12] Jami'in M A, Sutrisno I and Hu J 2012 Lyapunov learning algorithm for Quasi-ARX neural network to identification of nonlinear dynamical system. *IEEE International Conference on Systems, Man, and Cybernetics (SMC)* (Seoul, 2012) pp 3147-3152 doi: 10.1109/ICSMC.2012.6378275
[13] Jami'in M A, Sutrisno I and Hu J 2013 Deep searching for parameter estimation of the linear time invariant (LTI) system by using Quasi-ARX neural network. *The 2013 International Joint Conference on Neural Networks (IJCNN)* (Dallas-TX, 2013) pp 1-5 doi: 10.1109/IJCNN.2013.6707091
[14] Jami'in M A, Sutrisno I and Hu J 2015 Maximum power tracking control for a wind energy conversion system based on a quasi ARX neural network model. *IEEJ Trans Elec Electron Eng* **10** pp 368-375
[15] Jami'in M A, Sutrisno I, Hu J, Mariun N B and Marhaban M H 2014 An adaptive predictive control based on a quasi-ARX neural network model. *13th International Conference on Control Automation Robotics & Vision (ICARCV)* (Singapore, 2014) pp 253-258 doi: 10.1109/ICARCV.2014.7064314
[16] Sutrisno I, Jami'in M A and Hu J 2014 Modified fuzzy adaptive controller applied to nonlinear systems modeled under quasi-ARX neural network. *Artif Life Robotics* **19** pp 22–26 https://doi.org/10.1007/s10015-013-0137-6
[17] Sutrisno I, Jami'in M A and Hu J 2014 An improved elman neural network controller based on quasi ARX neural network for nonlinear systems. *IEEJ Trans Elec Electron Eng* **9** pp 494-501
[18] Sutrisno I, Che C and Hu J 2014 An improved adaptive switching control based on quasi-ARX neural network for nonlinear systems. *Artif Life Robotics* **19** pp 347–353
[19] Sutrisno I, Jami'in M A and Hu J 2012 Neural predictive controller of nonlinear systems based on quasi-ARX neural network. *18th International Conference on Automation and Computing (ICAC)* (Loughborough, 2012) pp 1-6
[20] Widodo R B et al. Grasping and attached mode in human computer interaction in the study of mouse substitution. *Journal of Physics: Conference Series* **1196** 1 012034
[21] Sutrisno I, Jami'in M A, Hu J and Marhaban M H 2015 Self-organizing quasi-linear ARX RBFN modeling for identification and control of nonlinear systems. *54th annual conference of the society of instrument and control engineers of Japan (SICE)*. IEEE
[22] Jami'in M A, Sutrisno I and Hu J 2014 Nonlinear adaptive control for wind energy conversion systems based on quasi-ARX neural networks model. *Procedings of the International Multi-Conference of Engineers and Computer Scientists* (March 12-14, 2014) (Hong Kong)
[23] Sutrisno I 2009 *Pemrograman Komputer dengan Software Matlab* (Surabaya: ITS Press)

IOP Conf. Series: Materials Science and Engineering 1175 (2021) 012006 doi:10.1088/1757-899X/1175/1/012006

Design floating robot of shallots irrigation with GPS based and using the waypoint navigation method

F Nofandi[1], R H Devandra[2], S Hasugian[3], I Sutrisno[4] and E Setiawan[5]
[1,3]Politeknik Pelayaran Surabaya, Indonesia
[2,4,5]Politeknik Perkapalan Negeri Surabaya, Indonesia

E-mail: farisnofandi@gmail.com

Abstract. Shallots are tuber plants that are used as a spice in cooking. Watering shallots in the right way regularly will produce quality shallots. Currently many shallot farmers in Indonesia still need the services of daily workers (laborers), especially farmers in Kab. Nganjuk, East Java. The watering process is still manual. In addition, it is difficult to find daily workers (laborers). This of course will create losses for farmers. With technological developments in the current era, the writer wants to make a tool that can be used by shallot farmers. This tool uses GPS to navigate the trench traversed using the waypoint method. Apart from using GPS, this tool also uses ultrasonic sensors which function to keep the robot in the trench line. For watering this robot uses a dc water pump with high pressure so that it can reach the plants from the trench. This tool can be operated using an android application that makes it easier for farmers.

1. Introduction

The development of technology in the current era can be said to be very fast. There are many studies to develop technology in various industrial sectors, one of which is in the agricultural industry sector. Our country, Indonesia, has enormous agricultural potential. One of them is shallot farming.

Shallots are tuber plants that are used as a spice in cooking. This plant is very much we find in various regions in Indonesia. Cultivating shallots is very profitable if done properly and correctly. Planting shallots must be paid close attention to and need good care in order to produce quality shallots. One important treatment is regular watering of the onion plants. The way to water onions is to make a trench between the beds where the onion plants grow, which functions to store water that will be poured on the shallot plants. Most farmers in Indonesia, especially in the Nganjuk Regency, East Java, currently still depend on daily workers (laborers) to carry out maintenance, especially shallot farmers. During the dry season when at that time the price of shallots can go up higher, but it is inversely proportional to the cost of maintenance and labor because it takes longer. Due to the watering process is still manual and it is difficult to find daily work (laborers). Things like this can make farmers lose and pay more [1].

From the above background, the researchers made and designed a final project entitled "Design Floating Robot of Shallots Irrigation With GPS Based and Using the Waypoint Navigation Method" to solve the problems of farmers in the difficulty of finding daily laborers. This robot uses a trench as a passage as well as a source of water to be sprayed on onion plants and is based on android.

Content from this work may be used under the terms of the Creative Commons Attribution 3.0 licence. Any further distribution of this work must maintain attribution to the author(s) and the title of the work, journal citation and DOI.
Published under licence by IOP Publishing Ltd

IOP Conf. Series: Materials Science and Engineering 1175 (2021) 012006 doi:10.1088/1757-899X/1175/1/012006

2. Methodology

2.1. System Design

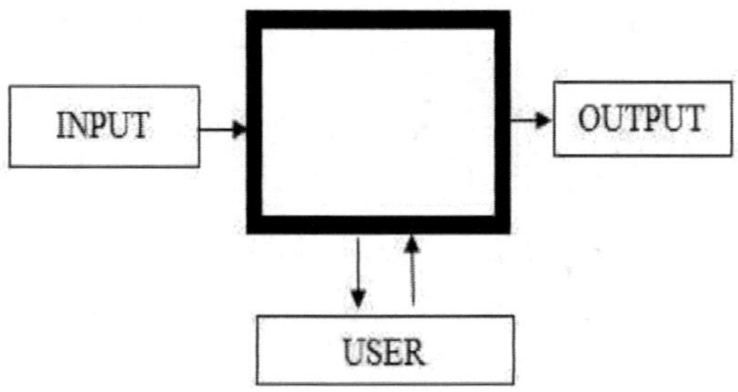

Figure 1. System Concept.

Based on Figure 1 this system design, it is explained that in a system there is a system block concept that is used to design and create a system. In the system block concept, it will be explained that in a system there are 2 inputs in the form of distance and coordinate points. The two inputs send data from the robot travel process which is then sent and processed to the Arduino microcontroller. Furthermore, the results of data processing are forwarded to the actuator control process as the output of this system. Then to make it easier for operators to monitor, the system is integrated with an interface in which there is an automatic actuator monitoring and control system.

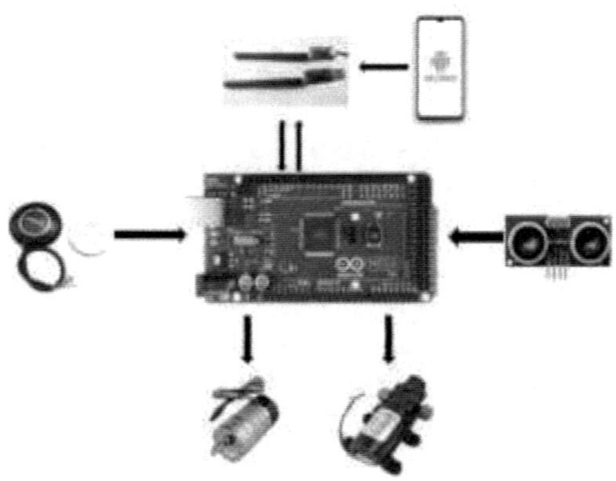

Figure 2. Design Hardware.

In Figure 2 it is explained that when the robot is placed in the trench, then the ultrasonic sensor as a distance sensor detects the distance between the trench wall and the robot so that the robot does not hit the track. Furthermore, the GPS Module will read the coordinates of the robot if it is as desired. The results of the data are then sent to the Arduino microcontroller. Then the results of the detection of all inputs will turn on and off the water pump, and the dc motor is in accordance with the results of input processing. All data and conditions that are monitored and controlled will be entered and displayed on Android.Based on the explanation in Figure 2 above, here are the components needed:

2.1.1.Arduino Mega.

Arduino Mega is an Arduino board that uses an ATmega 2560 microcontroller IC. This board has 54 digital I / O pins, 15 of which can be used as PWM PIN outputs, 16 analog inputs, 4 UARTs. Arduino

IOP Conf. Series: Materials Science and Engineering 1175 (2021) 012006 doi:10.1088/1757-899X/1175/1/012006

here is the brain of this final project prototype. Because in the Arduino Mega, data obtained from all sensors and control methods are processed in such a way and will instruct the actuators to carry out their duties according to the input, so that the set point of the system being built is achieved.

Figure 3. Arduino Mega 2560.

2.1.2.Telemetry.
Telemetry is a process of measuring an object whether space, object, nature or other parameters, 26 when the required results are obtained it will be immediately sent to a place with a certain distance using either wired or wireless media (without usingcables) using the data transmission method.

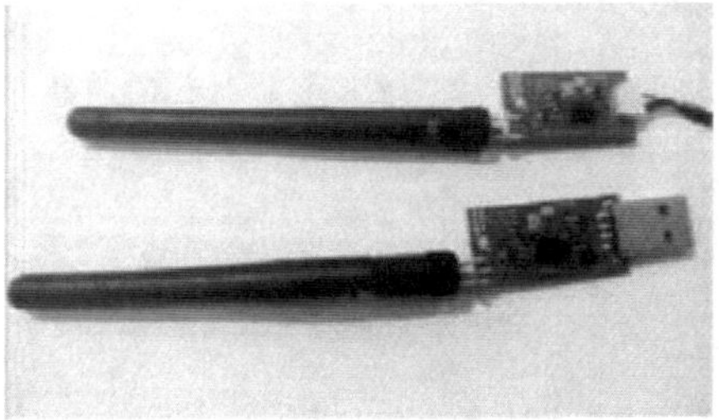

Figure 4. Telemetry.

2.1.3.Waypoint Method.
Making waypoints can be done by determining the longitude and latitude points in the application that has been made later, if 2 or more waypoints have been formed, draw a straight line to connect the waypoints. Then after the formation of the waypoints, the predetermined data will be sent to the Arduino in the robot.

2.1.4.DC Motor 12V.
DC motor is one type of electric motor that is often used today. DC motors require a direct voltage supply to the field coil to be converted into mechanical energy.

Figure 5. DC Motor.

2.1.5. Water Pump 12v.

Water Pump is a mechanical device used to raise fluids from lowlands to highlands or to flow fluids from low pressure areas to high pressure areas and also as a flow rate booster in a piping network system. In principle, the pump converts the motor mechanical energy into fluid flow energy. The energy received by the fluid will be used to increase the pressure and overcome the resistance found in the channel through which it is passed.

Figure 6. Water Pump.

2.1.6. GPS Ublox Neo M8N.

In this system, GPS is very influential because all system processes depend on the accuracy of the value provided by GPS. The Neo M8N GPS module is very easy to connect to Arduino. This GPS can work with an input voltage of 3VDC. This GPS is used to bring up the coordinates of the robot.

Figure 7. GPS Ublox M8N.

2.1.7. Ultrasonic Sensor HC-SR04.

In this system, the sensor used is an ultrasonic sensor. Here the proximity sensor functions to help the robot stay on track. This sensor can read distances of up to 4 meters with a minimum distance of 2 cm which is very sufficient for the needs of this study. This sensor can work with an input voltage of 5VDC.

Figure 8. Ultrasonic Sensor.

2.2. The System Works

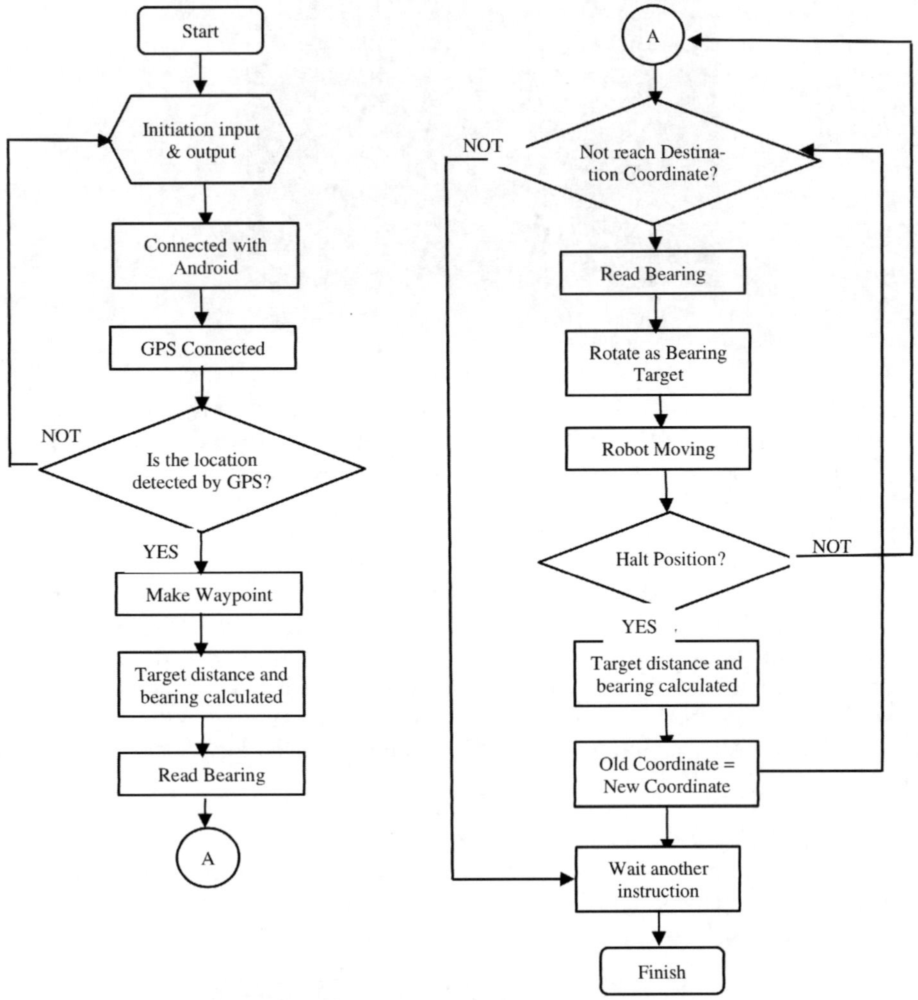

Figure 9. Flowchart System.

Figure 9 is the working principle used from this final project. In a system that is designed, GPS is an important part of the system because all system processes depend on the accuracy of the values provided by GPS. Android devices are not only a monitoring tool but also a navigation system calculation as input for the robot to move autonomously.

Figure 10. Floating Robot Design.

In this system, the actuator used is a 12v DC motor with 2 motors to drive the propeller of this floating robot, the next actuator is a 12v dc water pump with 80Psi water spray pressure. This actuator is used to spray water from the trench to the shallot plants.

3. Results and Discussion

3.1.Floating Robot Mechanical Testing

The mechanical testing process is carried out based on the realization of a pre-designed design. Testing is carried out after the floating robot is formed, this stage is the first step in making the robot, to determine the condition of the floating robot, whether it is suitable for use or needs repair before use. One indicator of the feasibility of a floating robot is that there is no leak, so that the floating robot can float in the water perfectly. The next stage, the robot is given a load according to the components used to determine the strength when it is in the water and see the movement performance with the maximum load carried.

3.2.Ultrasonic HC SR04 Sensor Testing

The HC-SR04 ultrasonic sensor test was conducted to determine the measurement data. In testing the HC-SR04 sensor using a comparator in the form of a ruler where this ruler is the reference for the actual distance data, as in Figure 11.

Figure 11. Ultrasonic Sensor Testing.

Testing is done by placing the object in front of the sensor then the distance is adjusted to 5 times the test. Then after that it is compared to the output on the Arduino serial monitor with a ruler.

Below is table 1 comparison of the ultrasonic sensor readings with a ruler comparison.

Table 1. Comparison of Ultrasonic Sensor.

No.	Arduino (in cm)	Ruler (in cm)	Error (%)
1.	3	3	0
2.	5	5	0
3.	10	10	0
4.	15	15	0
5.	20	20	0
6.	25	26	3,85
	Average Error		0.64

Table 1 shows the analysis of the HC-SR04 ultrasonic distance sensor testing using Arduino. The sensor is able to read the distance with an average error value of 0.64%.

3.3. GPS Testing

This GPS test is used to determine the position with coordinate points and serves as a sensor to determine the distance to which the robot moves. To prove the accuracy of the ublox M8N gps module, the Google Maps application is used as a gps validator. Google Maps will provide the smartphone location coordinate point data so that from the test results of the gps module coordinates with google maps points an error is obtained from the ublox M8N gps module.

Figure 12. GPS Testing.

From Figure 12 gps testing, the data received by Arduino are the coordinates of the GPS. In this test, the latitude and longitude points are obtained which are compared with the Google Maps application, here are the results of the GPS test.

Figure 13. Data of GPS Arduino.

Figure 14. Google Maps as a comparison of sensor.

Based on Figure 13 and Figure 14, it can be seen that the GPS reading on the Arduino is very stable and the coordinates displayed on the Arduino look the same as the points on google maps.

3.4. Android Interface

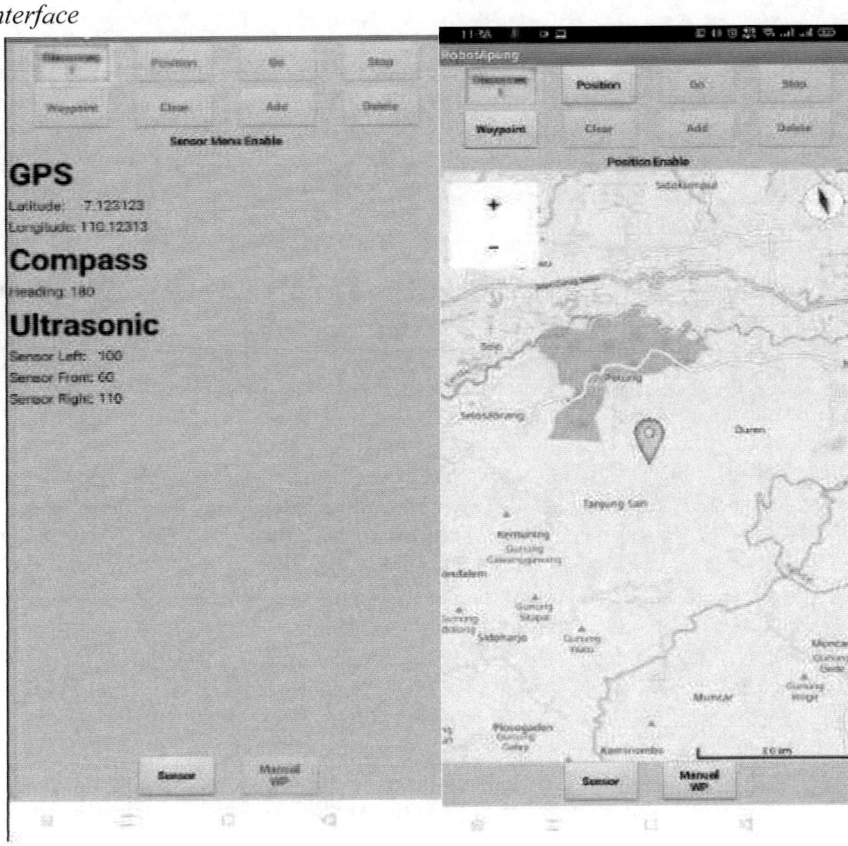

Figure 15. Android Interface.

Based on Figure 15 The function of this android is the control and monitoring of the android system used to select the waypoint points that the robot will go to so that the robot can water the plants according to the path. In this display, you can also see the data output from the sensor sensors used.

4. Conclusion

This floating robot for sprinkling onions is a robot that is able to navigate rice fields with trenches automatically with the help of a waypoint navigation system. In addition to using waypoint navigation, the robot also uses ultrasonic sensors so that the robot does not hit an obstacle and remains on the trench. From the realization and testing of the tools carried out, it can be concluded that the ublox m8n gps system is very stable in displaying coordinate points, the error obtained from the test is not more than less than 3 meters. In testing the hc-sr04 ultrasonic sensor, a very small error of 0.64% was found, so it can be concluded that this type of ultrasonic sensor is suitable as a robot safety so as not to hit trenches or robot lines.

5. References

[1] Muharom et al. 2019 Rancang bangun sistem penyiram tanaman bawang merah secara otomatis. *Prosiding Seminar Nasional Sains dan Teknologi Terapan*

[2] Hidayat et al. 2016 Pengembangan sistem navigasi otomatis pada UAV (Unmanned Aerial Vehicle) dengan GPS (Global Positioning System) Waypoint. *Jurnal Teknik ITS* **5** 2

[3] Ichwan et al. 2013 Pembangunan prototipe sistem pengendalian peralatan listrik pada platform android. *Jurnal Informatika* **4** 1

[4] Nugroho et al. 2016 Sistem navigasi gerak roboboat berdasarkan GPS menggunakan metode Waypoint. *E-Proceeding of Engineering* **3** 2

[5] Nurrizki D B 2019 *Klasifikasi Pengukuran Level Gas pada Proses Pembuatan MSG dengan Menggunakan Neural Network.* (Tugas Akhir: Politeknik Perkapalan Negeri Surabaya)

[6] Pramana et al. 2018 Optimasi sistem kendali lampu lalu lintas cerdas menggunakan metode logika Fuzzy. *E-Proceeding of Engineering* **5** 3

[7] Saputra et al. 2018 Autonomous Surface Vehicle sebagai alat pemantau lingkungan menggunakan metode navigasi Waypoint. *Jurnal Teknik ITS* **7** 1

[8] Yuski et al. 2017 Rancang bangun jangkar motor DC. *Jurnal Berkala Saintek* **2**

[9] Jami'in M A, Hu J, Marhaban M H, Sutrisno I and Mariun N B 2016 Quasi ARX neural network based adaptive predictive control for nonlinear systems. *IEEJ Trans Elec Electron Eng* **11** pp 83-90

[10] Jami'in M A, Sutrisno I and Hu J 2012 Lyapunov learning algorithm for Quasi-ARX neural network to identification of nonlinear dynamical system. *IEEE International Conference on Systems, Man, and Cybernetics (SMC)* (Seoul, 2012) pp 3147-3152 doi: 10.1109/ICSMC.2012.6378275

[11] Jami'in M A, Sutrisno I and Hu J 2013 Deep searching for parameter estimation of the linear time invariant (LTI) system by using Quasi-ARX neural network. *The 2013 International Joint Conference on Neural Networks (IJCNN)* (Dallas-TX, 2013) pp 1-5 doi: 10.1109/IJCNN.2013.6707091

[12] Jami'in M A, Sutrisno I and Hu J 2015 Maximum power tracking control for a wind energy conversion system based on a quasi ARX neural network model. *IEEJ Trans Elec Electron Eng* **10** pp 368-375

[13] Jami'in M A, Sutrisno I, Hu J, Mariun N B and Marhaban M H 2014 An adaptive predictive control based on a quasi-ARX neural network model. *13th International Conference on Control Automation Robotics & Vision (ICARCV)* (Singapore, 2014) pp 253-258 doi: 10.1109/ICARCV.2014.7064314

[14] Sutrisno I, Jami'in M A and Hu J 2014 Modified fuzzy adaptive controller applied to nonlinear systems modeled under quasi-ARX neural network. *Artif Life Robotics* **19** pp 22–26 https://doi.org/10.1007/s10015-013-0137-6

[15] Sutrisno I, Jami'in M A and Hu J 2014 An improved elman neural network controller based on quasi ARX neural network for nonlinear systems. *IEEJ Trans Elec Electron Eng* **9** pp 494-501

[16] Sutrisno I, Che C and Hu J 2014 An improved adaptive switching control based on quasi-ARX neural network for nonlinear systems. *Artif Life Robotics* **19** pp 347–353

[17] Sutrisno I, Jami'in M A and Hu J 2012 Neural predictive controller of nonlinear systems based on quasi-ARX neural network. *18th International Conference on Automation and Computing (ICAC)* (Loughborough, 2012) pp 1-6

[18] Rinanto N and Khumaidi *Aplikasi Android-Rasspberry Pi pada Kapal Tanpa Awak untuk Pencarian Korban Kecelakaan Laut*

[19] Widodo R B et al. Grasping and attached mode in human computer interaction in the study of mouse substitution. *Journal of Physics: Conference Series* **1196** 1 012034

[20] Sutrisno I, Jami'in M A, Hu J and Marhaban M H 2015 Self-organizing quasi-linear ARX RBFN modeling for identification and control of nonlinear systems. *54th annual conference of the society of instrument and control engineers of Japan (SICE)*. IEEE
[21] Jami'in M A, Sutrisno I and Hu J Nonlinear adaptive control for wind energy conversion systems based on quasi-ARX neural networks model. *Procedings of the International Multi-Conference of Engineers and Computer Scientists* (March 12-14, 2014) (Hong Kong)
[22] Sutrisno I 2009 *Pemrograman Komputer dengan Software Matlab* (Surabaya: ITS Press)

IOP Conf. Series: Materials Science and Engineering 1175 (2021) 012007 doi:10.1088/1757-899X/1175/1/012007

Linear frequency modulation signal for channel impulse response measurement in shallow water environment

Y Widiarti[1], M B Rahmat[2], Adianto[3] and P Asri[4]

[1,2,3,4] Marine Electrical Engineering Department,
Shipbuilding Institute of Polytechnic Surabaya,
Surabaya, Indonesia

E-mail: yuning.widiarti@ppns.ac.id

Abstract. Understanding the characteristics of channel propagation plays a very important role in the design of underwater acoustic communication. Underwater acoustic channel (UAC) model is usually carried out by measuring directly at a specific location and time. This paper presents the use of Linear Frequency Modulation (LFM) signal for channel impulse response estimation and power delay profile (PDP) measurement in a shallow water environment. Measurements were made in a laboratory-scale towing tank with three variations of distance and depth. Furthermore, the measurement results are compared with the existing underwater acoustic channel model. The measurement results show that the response impulses at the three distances have almost the same pattern and the resulting amplitude is similar to the existing underwater acoustic channel model simulation. The RMS delay spread generated from the three distances has a value of about 0,2 msec. Meanwhile, the maximum excess delay obtained at a distance of 120 m has the highest value, which is 29 msec.

1. Introduction

The role of underwater communication and network is crucial for commercial and military needs. In recent decades, the amount and quality of research in the field of underwater communication has risen rapidly. The need to communicate between sensor nodes in a sensor network requires the characterization of an underwater acoustic channel. Due to the complexity of the underwater environment for communication, many researchers have focused on different aspects of characterizing an underwater acoustic communication channel as in [1]. Each environment has different characteristics that affect the performance of a digital communication system [2]. Therefore, it is necessary to characterize the underwater communication channel based on the measurement of the channel parameters where the communication system is applied.

Sound propagation in the ocean is very complex and must be well understood. Moreover, sound propagation in a shallow water environment. Shallow water acoustic communication channel categorized as a multipath fading channel. Usually, in shallow water the signal experiences a long

Content from this work may be used under the terms of the Creative Commons Attribution 3.0 licence. Any further distribution of this work must maintain attribution to the author(s) and the title of the work, journal citation and DOI.
Published under licence by IOP Publishing Ltd

multipath delay spread, resulting in intersymbol interference (ISI) if the delay spread exceeds the symbol period in the communication system. Significant Doppler distribution in the frequency domain or short coherent time in the time domain is a phenomenon that often occurs on this channel. Due to reflections on the ocean's boundaries and refractions due to the sound speed that varies with depth, sound tends to propagate through multiple trajectories. Sea temporal variability combined with low sound speed can cause significant Doppler shifts. As a result, the channel will be affected by time and/ or frequency dispersion.

Some studies regarding the characterization of shallow water acoustic channels include the following: Loubet and Jourdain [3] who examined the Northern Sea at a depth of 500 m, Cook and Zaknich[4] at a depth of 4 m of Fremantle Fishing Boat Harbor in Western Australia. In 2011, Borowski conducted his research on multipath underwater acoustic communication channel[5] and P. Van Walree [6] in 2009. Research on probe signal for estimating impulse response has also been carried out in [7] using pseudo-noise (PN)-sequence signal, exponentially sine swept (ESS), and amplitude-modulated ESS (AM-ESS). The measurement of response impulse is to get the characteristics of the underwater acoustic channel.

This paper presents an evaluation of the use of Linear Frequency Modulation (LFM) signal as a probe signal for impulse response measurement in shallow water acoustic channel. In this study, the shallow water environment is represented by the environment of a towing tank which has dimensions of 200 m x 12 m x 6 m. The measurement data is used to get the propagation parameter, namely the channel impulse response (CIR). Furthermore, the results obtained are compared with the underwater acoustic channel (UAC) model performed by Parastoo Qarabaqi [8].

This paper is organized as follows. Section 2 describes the measurement setup, linear frequency modulation signal as a probe signal, and channel impulse response (CIR) estimation. Section 3 presents a comparison of the CIR obtained from measurement with the UAC model performed by Parastoo Qarabaqi. Section 4 summarizes the results that were obtained in this study.

2. Measurement-Based Model of Shallow Water Environment

2.1. Measurement Setup

The impulse response measurement is carried out on a towing tank belonging to the Indonesian Hydrodynamics Laboratory (IHL)-BPPT which has 200 m in length, 12 m in width, and 6 m in depth. Before the impulse response measurement process, the same scenario was carried out to determine the environmental noise character as in the study [9]. The measurements were made along with the towing tank with 3 variations of distance and depth. The scenario can be shown in Figure 1. The towing tank has a wall and bottom made of concrete material. The media inside is freshwater and with 999,9720 kg / m3 ~ 1,0 kg/liter density at 4°C (39°F) and the liquid has a temperature between 27°C-29°C. There are no transient waves and noise in the towing tank [10].

At figure 1, an underwater speaker is placed at a depth of 3 m below the surface as a sound source. There is a distance of 20 m between the pool wall and the speaker. The underwater speaker is connected to a power amplifier and a personal computer (PC) as a signal generator. As a receiver, the hydrophone array is arranged vertically with a space of 0,045 m between the hydrophones. This distance is set based on the wavelength and the space of the transmitter to the receiver. Thus, it can be expressed as $d < \lambda_{min}/2$ where d is the distance between the hydrophones and λ_{min} is the minimum wavelength of the received signal. The value obtained is a safe distance from aliasing. The hydrophone array is connected to a digital mixer, which is a device for recording sound transmitted from the underwater speaker and then received by the hydrophone array connected to the receiving PC for signal processing. The probe signal is generated using software on a PC equipped with a sound card and power amplifier.

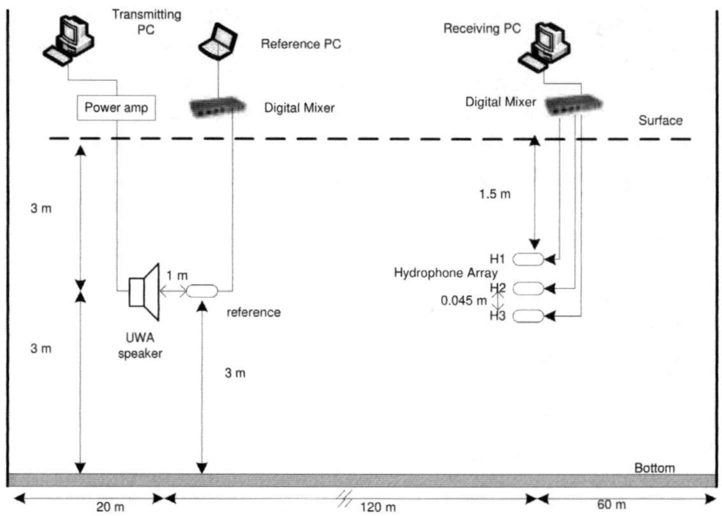

Figure 1. The arrangement of devices on towing tank

The conversion process to an acoustic signal is done by an underwater speaker device. The speaker is Aquasonic AQ339 with the following specifications: 135 Watt power, an impedance of 4 Ω, and a frequency between 20 Hz-17 kHz. The pattern of the speaker has almost omnidirectional with a peak at 0° angle [7]. Meanwhile, the UW hydrophone used has the following specifications: frequency range between 1 Hz to 100 kHz, sensitivity −190 dBre: 1 V/μPa(+4 dB, 20 Hz–4 kHz), and has an internal capacitance of 25 nF. The hydrophone reception pattern is almost omnidirectional and has the strongest response to the signal coming at an angle of 0° with a normal line.

2.2. Linear Frequency Modulation Signal as Probe Signal

The impulse response estimation is obtained by sending a probe signal from the transmitter to the receiver. In this measurement, an LFM signal, also known as chirp, is used as the probe signal. The determination of the LFM signal is based on the autocorrelation properties of the LFM signal with the following equation:

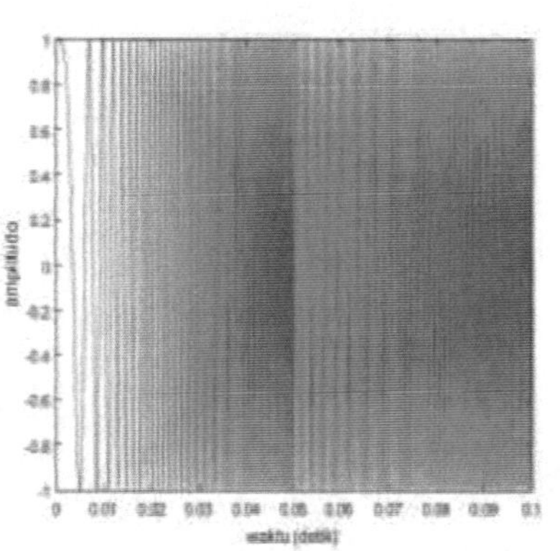

Figure 2. Chirp signal: (a) Time-domain

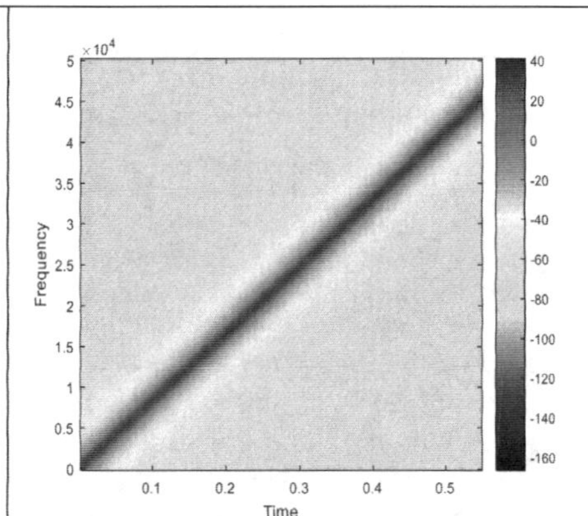

Figure 2. Chirp signal: (b) Spectogram of chirp

the following equation:

$$x(t) = exp\,(j\theta(t)) \quad (1)$$

θ(t) denotes the instantaneous phase with the following equation:

$$\theta(t) = 2\pi(f_0 t \pm Kt^2), \text{with} -\frac{T}{2} \le t \le \frac{T}{2} \quad (2)$$

f_0 denotes the center frequency at time t = 0, while K is the increase in frequency or chirp rate. Figure 2 shows the chirp signal in a time domain and its spectrum in the frequency domain.

Figure 2a shows an exponential chirp waveform, which is a sine wave that increases in frequency exponentially over a period of time. Figure 2b is a chirp signal spectrogram showing the linear level of frequency changes in the time function. The intensity of plotting is proportional to the energy content in the signal at the indicated frequency and time. LFM signal has been widely used as a channel probe signal in underwater acoustic communication because it has good autocorrelation properties [9]. In this measurement, the LFM signal used has a frequency range from 12 kHz to 15 kHz with a recording duration of 100 msec.

2.3. Impulse Response Estimation

The channel impulse response estimation is obtained by cross-correlating the LFM signal on the reference with the

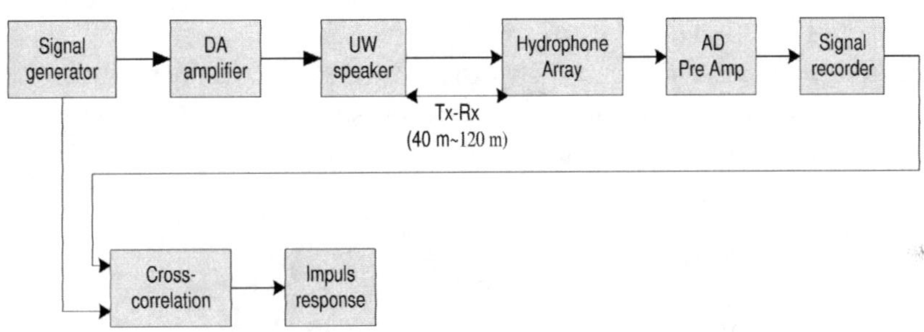

Figure 3. CIR estimation process

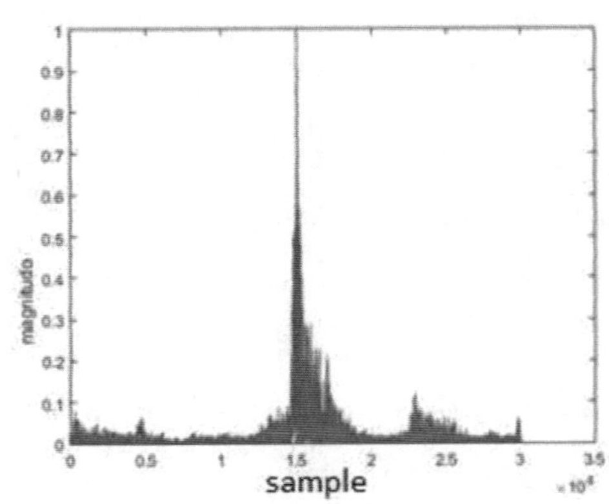

Figure 4. Cross-correlation of reference signals and received signals

LFM signal received on the hydrophone array. The impulse response estimation obtained by cross-correlation is shown in the block diagram of Figure 3. A reference signal is needed in the impulse response estimation process because the reference signal is a sample that is considered accurate with a distance of 1 meter from the source. There are three variations in this measurement scenario. Three variations of the distance between the source (transmitter) and receiver are: 40 m; 70 m; and 120 m and each distance varies in depth: 1,5 m; 2,5 m; and 3,5 m.

The probe signal is generated using software, and then the generated signal is transmitted to the receiver using an underwater speaker. Before being processed, the LFM signal received on each hydrophone must be synchronized with the reference signal first. This measurement uses three hydrophones on the receiver. The synchronization process aimsec to match the signal position in each frame. The next process is cross-correlating between the signal received and the reference signal. This process result can be shown in Figure 4. Then, a filtering process is carried out to get the impulse response estimation.

3. Comparison of The Underwater Acoustic Channel Model with The Measurement Result

The channel impulse response amplitude is obtained from the average amplitude of the recorded signal on each hydrophone. The impulse response is modeled as:

$$h(t) = \sum_{i=1,2,..} \sum_{j=1}^{k} A_{ji} \delta(t - \tau_{ji}) \quad (3)$$

where A_{ji} is attenuation caused by transmission loss and accumulation of the reflection coefficient and δ is the Dirac function. One method of measuring CIR can be done by autocorrelation, by utilizing a wideband test signal [7]. The value of -20 dB relative to the highest level of the impulse response signal is used as the lowest threshold level that will be processed in the analysis of the measurement results. The measurement results show that at all variations of distance, the signal experiences several reflections. CIR estimated is compared with the UAC model conducted by Parastoo-Qarabaqi [8]. The parameters used for the simulation are shown in Table 1.

Table 1. Simulation Parameter

Parameter	Value
Towing tank dimension	12 m x 200 m x 6 m
Tx-Rx distance	40 m; 70 m; 120 m
Depth variation	1,5 m; 2,5 m; 3,5 m
Sound speed	1500 m/sec
Minimum frequency	12 kHz
Sampling frequency	48 kHz
Bandwidth	3 kHz
Water density	1000 gram/cm^3
Chirp duration	100 msec
Doppler spread	4 Hz

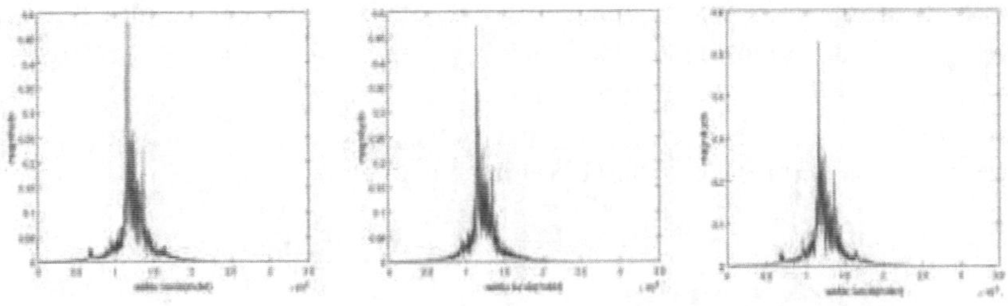

a b c

Figure 5. CIR at a distance of 40 m with a depth of (a) 1.5 m (b) 2.5 m (c) 3.5 m

Table 2. Parameter of PDP at 40 m

PDP parameter	Depth		
	1,5 m	2,5 m	3,5 m
Mean excess delay	11,54 msec	11,83 msec	11,835 msec
RMS delay spread	0,19 msec	0,22 msec	0,215 msec
Maximum excess delay	13,51 msec	11,45 msec	11,45 msec

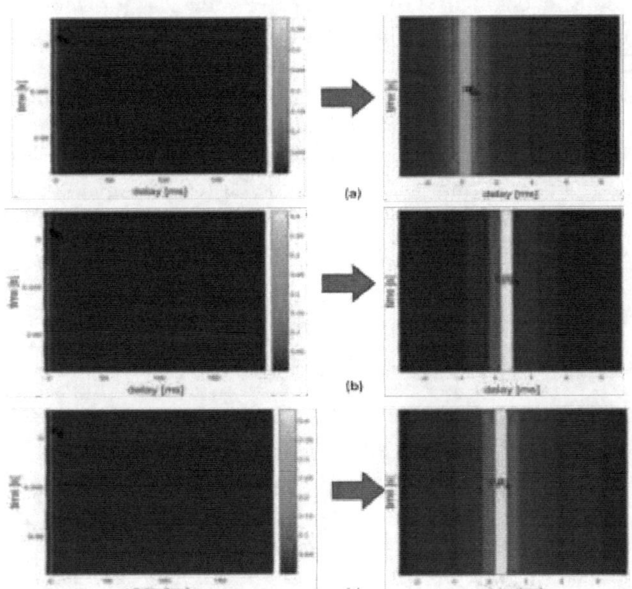

Figure 6. CIR of UAC Model at 40 m (a) 1.5 m (b) 2.5 m (c) 3.5 m

The channel impulse response (CIR) consists of a direct path and multipath signal. Figure 5 shows CIR at a distance of 40 m. The pattern has peak values at t = 11 msec and the sidelobes decrease regularly at t < 11 msec and t >11 msec, and at t > 20 msec the amplitude of the multipath is close to zero. The mean excess delay and the maximum excess delay value are relatively small. The amplitudes of the multipath produced are almost uniform at all depth variation.

Power Delay Profile (PDP) describes the average power as a function of time delay and different observation time and can be expressed in the following equation:

$$P(\tau) \approx \frac{1}{T}\int_{t-\frac{T}{2}}^{t+\frac{T}{2}} |y(t,\tau)|^2 dt \tag{4}$$

T in the equation is the observation time interval, and $|y(t,\tau)|^2$ denotes the square of the incoming signal at the receiver. Parameters related to PDP are the mean excess delay, RMS delay spread, and excess delay spread. The mean excess delay is the first moment of PDP with the following equation:

$$\bar{\tau} = \frac{\sum_k P(\tau_k)\tau_k}{\sum_k P(\tau_k)} \tag{5}$$

The RMS delay spread is at the root of the second PDP moment and is defined in the following equation:

$$\sigma_\tau = \sqrt{\overline{\tau^2} - (\bar{\tau})^2} \tag{6}$$

with:

$$\overline{\tau^2} = \frac{\sum_k P(\tau_k)\tau_k^2}{\sum_k P(\tau_k)} \tag{7}$$

Maximum excess delay (X dB) is the delay time on multipath, in which the energy of the incoming signal at the receiver still exceeds the lowest limit (≥ -20dB). The power delay profile parameter has the values as shown in Table 2.

The underwater acoustic channel (UAC) model is time-varying and the model is useful for statistical observation of the underwater channel in a multipath environment. Figure 6 shows the simulation result of CIR at a distance of 40 m. When compared with the measurement results in Figure 5, there are similarities in the resulting multipath patterns. The multipath that occurs appears denser at $t \geq 0$ msec, meanwhile, the lower amplitude of the trajectory is getting to the right. This can be observed in Figure 6 on the right which is an enlargement of the image to the left and the yellow line shows a higher amplitude value than the blue colour. The largest multipath amplitudes are generated at a depth of 2,5 m and 3,5 m from the surface, while the measurement results have a slightly larger amplitude achieved at a depth of 3,5 m.

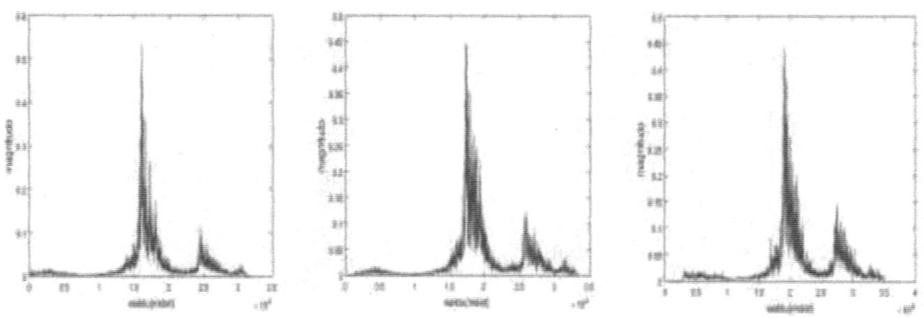

a b c)

Figure 7. CIR at a distance of 70 m with a depth of (a) 1,5 m (b) 2,5 m (c) 3,5 m

Table 3. Parameter of PDP at 70 m

PDP parameter	Depth		
	1,5 m	2,5 m	3,5 m
Mean excess delay	16,37msec	17,57msec	19,24msec
RMS delay spread	0,2msec	0,285msec	0,24msec
Maximum excess delay	24,54msec	26,1msec	27,97msec

At 70 m, the resulting multipath pattern is slightly different when compared to the multipath pattern at 40 m. This can be shown in Figure 7. The peak amplitude values are obtained at t = 16 msec at a depth of 1,5 m, while at a depth of 2,5 m and 3,5 m the peak values are obtained at t = 17 msec and t = 19 msec, respectively. At this distance, there are several groups of multipath, between the value of t = 16 msec to about t = 25 msec. There is an increase in the amplitude at t > 25 msec and forming a multipath pattern like in the first group, but with a smaller amplitude. Mean excess delay and the maximum excess delay produced are longer when compared with a distance of 40 m. Detail of the power delay profile (PDP) parameter at a distance of 70 m can be seen in Table 3.

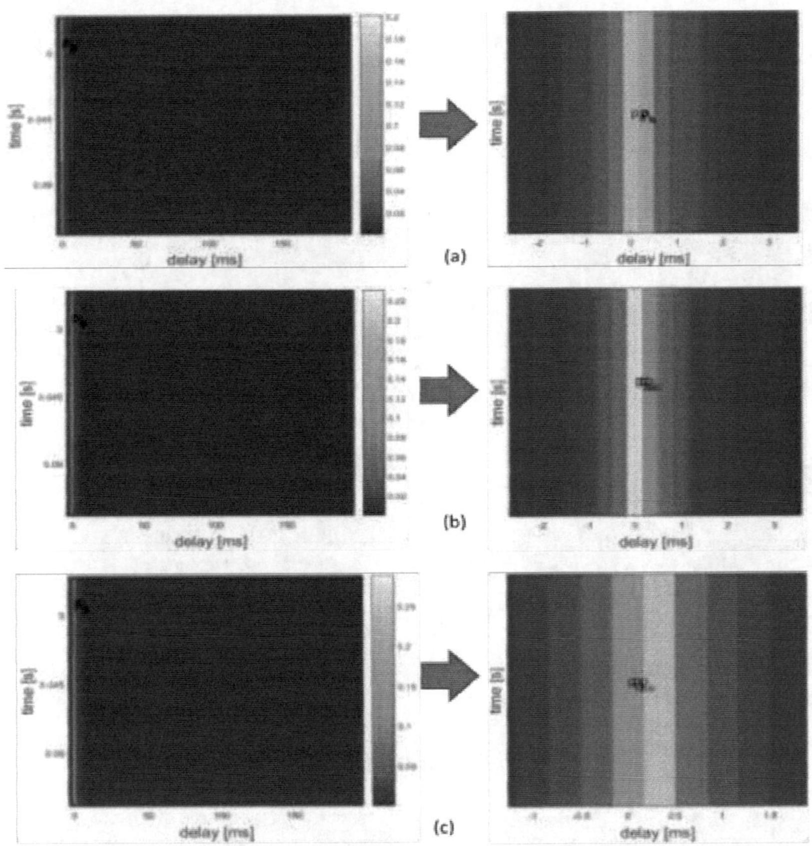

Figure 8. CIR of UAC Model at 70 m (a) 1.5 m (b) 2.5 m (c) 3.5 m

The amplitude of the multipath produced is almost the same as the amplitude at a distance of 40 m. When compared with the simulation results in Figure 8, the resulting multipath pattern shows the density at $t \geq 0$ msec. This is similar to the multipath pattern generated at a distance of 40 m.

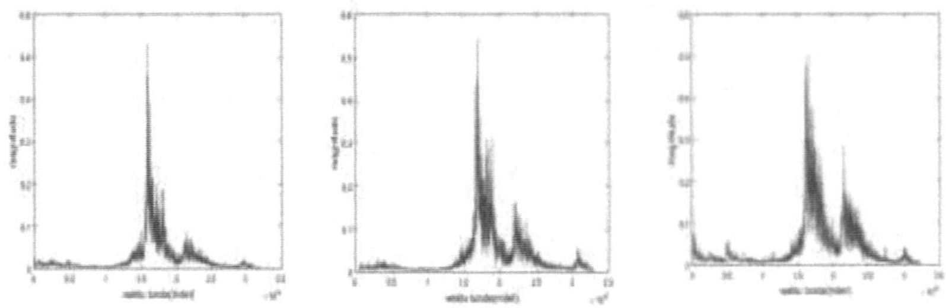

a b c

Figure 9. CIR at a distance of 120 m with a depth of (a) 1.5 m (b) 2.5 m (c) 3.5 m

Table 4. Parameter of PDP at 120 m

PDP parameter	Depth		
	1,5 m	2,5 m	3,5 m
Mean excess delay	18,17 msec	16,72 msec	15,995 msec
RMS delay spread	0,34 msec	0,27 msec	0,24 msec
Maximum excess delay	28,89 msec	27,23 msec	29 msec

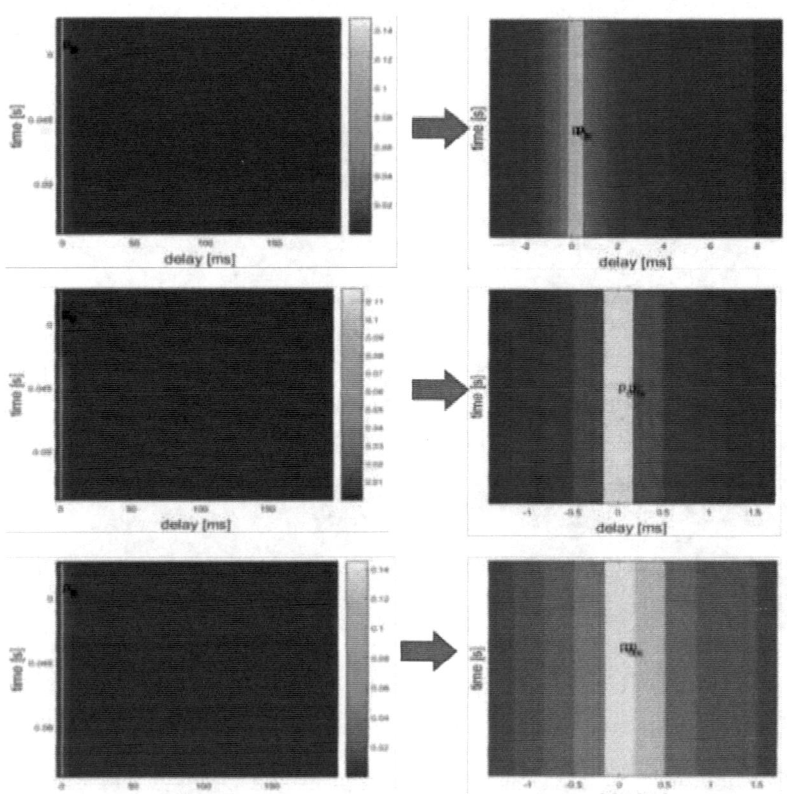

Figure 10. CIR of UAC Model at 120 m (a) 1.5 m (b) 2.5 m (c) 3.5 m

However, at a distance of 70 m, the amplitude of the multipath is smaller when compared to the amplitude at a distance of 40 m. This is because the path losses generated at longer distances will be greater[11].

The resulting impulse response at a distance of 120 m can be shown in Figure 9. This multipath pattern is similar to the resulting impulse response at a distance of 70 m where there are several groups of multipath. The first group has a pattern similar to the pattern at a distance of 40 m with peak amplitude values obtained at t = 16 msec at a depth of 1.5 m and t = 17 msec at a depth of 2.5 m and 3.5 m. The second group has almost the same pattern as the first group with a smaller amplitude. The resulting mean excess delay and maximum excess delay are shorter than the 70 m distance. From the three variations of the distance, the result of the impulse response obtained has a relatively constant amplitude with a slightly different path pattern between the multipath obtained at a distance of 70 m and 120 m. If compared with the simulation results in Figure 10, there is a similarity in the resulting track pattern, which is denser when $t \geq 0$ msec. In the simulation results, the amplitude of the multipath tends to decrease with increasing distance. However, the real measurement results at a distance of 120 m tend to be constant and even higher when compared to the amplitude at a distance of 40 m. This is due to the reflection on the pool wall or the surface and surface of the pool which allows there to be a point where the reflections that occur will reinforce each other.

4. Conclusion

The measurement evaluation of CIR in the shallow water environment has been conducted using three variations of the distance between transmitter and receiver: 40 m, 70 m, and 120 m with three depths: 1,5 m, 2,5m, dan 3,5m. From the whole series of measurements, the results show that the channel impulse response at three distance variations has almost the same multipath pattern. LFM or chirp signal can be used to estimate channel impulse response well. The PDP parameter shows that the RMS delay spread generated at the three distances has an average value of 0.2 msec. While the highest maximum excess delay is 29 msec obtained at a distance of 120 m with a depth of 3.5 m.

References

[1] Eggen T Baggeroer A and Preisig J 2000 Communication over doppler spread channels Part i: channel and receiver presentation *IEEE Journal of Oceanic Engineering*. **25** 1 pp 62–71
[2] Borowski B 2009 Characterization of a very shallow water acoustic communication channel in OCEANS 2009 *MTS/IEEE Biloxi-Marine Technology for Our Future: Global and Local Challenges*. p 11
[3] Loubet G and Jourdain G 1993 Characterization of the underwater medium as an acoustical horizontal transmission channel in *Proc. of the IEEE International Conference on Acoustics, Speech and Signal Processing*. **1** pp 329-332 April 1993
[4] Cook G and Zaknich 1998 A chirp sounding the shallow water acoustic channel in *Proc. of the IEEE International Conference on Acoustics, Speech and Signal Processing*. **4** pp 2521-2524 May 1998
[5] Borowski B 2009 Characterization of a very shallow water acoustic communication channel in OCEANS 2009 *MTS/IEEE Biloxi-Marine Technology for Our Future: Global and Local Challenges*. p 110
[6] Van Walree P 2011 Channel sounding for acoustic communications: techniques and shallow-water examples *Norwegian Defence Research Establishment (FFI), Tech. Rep. FFI-rapport*. **7**

[7] Santoso T B, Widjiati E, Wirawan and Hendrantoro G 2016 The evaluation of probe signals for impulse response measurements in shallow water environment in *IEEE Transactions on Instrumentation and Measurement.* **65** 6 pp 1292-1299 June 2016. doi: 10.1109/TIM.2016.2519768
[8] Qarabaqi, P and Stojanovic, M 2013 Statistical characterization and computationally efficient modeling of a class of underwater *IEEE Journal of Oceanic Engineering.* **38** 4 pp 701–717
[9] Widiarti, Y and Suwadi, W 2019 Experimental measurement of time-reversal-OFDM technique for underwater acoustic communication in the presence of Gaussian noise in *Proc. 2019 International Conference on Information and Communications Technology (ICOIACT).* pp 297–301 Yogyakarta, Indonesia
[10] Widiarti, Y, Wirawan and Suwadi 2020 Joint time-reversal precoding and spatial diversity technique for acoustic communication in shallow water environment *International Journal of Intelligent Engineering and Systems.* **13** 1 pp 237-247
[11] Stojanovic, M and Preisig, J 2009 Underwater acoustic communication channels: propagation models and statistical characterization *IEEE Communications Magazine.* **47** 1 pp 84-89

Acknowledgments

The author would like to thank IHL and all the crew for helping to make this measurement run smoothly and Shipbuilding Institute of Polytechnic Surabaya which has provided full support through the 2020 DIPA fund.

IOP Conf. Series: Materials Science and Engineering 1175 (2021) 012008 doi:10.1088/1757-899X/1175/1/012008

Fish aggregating devices as buoyant apparatus for improving safety on traditional fishing vessel

P A Wibawa[1,4] , A Wahidin[2] and L Subiyanto[3]
[1, 2] Shipbuilding Engineering Department, Politeknik Perkapalan Negeri Surabaya, Indonesia
[3] Ship Electrical Engineering Department, Politeknik Perkapalan Negeri Surabaya, Indonesia

E-mail: putuarta@ppns.ac.id

Abstract. The availability of standard safety equipment on fishing vessels in Indonesia is still a problem that needs to be addressed. Liferaft, which functions as a rescue raft for the crew when the ship has an accident and sinks, is rarely found on traditional fishing vessel in Indonesia. Traditional fishermen believed that some of the equipment on board can be used as safety equipment, such as the floating part of Fish Aggregating Devices. Proper buoyant apparatus has been designed to fulfil the minimum standard for safety equipment according to FAO/ILO/IMO recommendation. The designed buoyant apparatus has other function as a floating part for FAD. The buoyant apparatus can hold 6 crew in the water, and at the same time can support FAD module that consist of 1 unit 100WP solar panel, 4 unit batteries 12v 18Ah, and 4 light @10watt.

1. Introduction

Lack of official data regarding number of accidents per year in fisheries sector has caused difficulties to have a complete overview about the condition of safety aspect in fisheries sector in Indonesia [1]. The data that have been provided by The People's Coalition for Fisheries Justice (KIARA) can be used as rough description about the implementation of safety in fisheries sector in Indonesia. According to KIARA, there were 186 fishers who died at sea in 2012, as many as 225 in 2013 and 210 people in 2014 [2].

The above data confirms that the awareness of fishermen in Indonesia to implement responsible fishing practices, as required by the Food and Agriculture Organization (FAO) in the Code of Conduct for Responsible Fisheries, is still very low [3]. The Food and Agriculture Organization (FAO) on the *Global Review of Safety at Sea in the Fisheries Sector 2018*, also identified that the main cause of accidents in Indonesian Water regarding the fishing vessels were poor construction of the boats, lack of safety awareness, and lack of safety equipment.

In fact, the availability of standard safety equipment is still very rarely found on fishing vessels in Indonesia. Although the availability of standard safety equipment has become a compulsory to get a Port Clearance Letter for fishing vessels that will leave the fishing harbour. However, by considering the economic condition of most fishermen in Indonesia, therefore, the enforcement of this regulation has become applied in a more flexible way for local fishing boats [4].

Content from this work may be used under the terms of the Creative Commons Attribution 3.0 licence. Any further distribution of this work must maintain attribution to the author(s) and the title of the work, journal citation and DOI.
Published under licence by IOP Publishing Ltd

Figure 1. FAD with simple floating device

Contrary to fishermen's understanding of the high risk of their work, the willingness of fishermen to equip their boats with safety equipment is still low. This is because fishermen assume the existence of safety equipment on board is less important when compared to equipment that is directly related to their work as fishermen such as fishing gears or hauling machines. Moreover, fishermen argue that many of standard safety equipment can be replaced by other tool or equipment available on their fishing vessels [4].

One of the safety equipment that must be available on board, but almost never found on fishing vessels in Indonesia is a liferaft. The results of interviews with local fishermen in Muncar, Banyuwangi conducted by Wibawa (2016) showed that fishermen considered that part of the Fish Aggregating Devices (FAD) for floating purposes could function as a liferaft. However, if you pay attention to the construction of the current FAD, the construction of floating structure that provide buoyancy to FAD is not suitable to be used as a Buoyant Apparatus for safety equipment, as shown in figure 1.

According to FAO on *The Safety Recommendations for Decked Fishing Vessels of Less than 12 metres in Length and Undecked Fishing Vessels,* every vessel should bring sufficient equipment for recovering persons from the water. The Table 1 shows Life-saving appliances for vessels of different design categories according to FAO [5].

Based on this Table 1, design category C and D of vessel are allowed to have buoyant apparatus to replace liferaft on board. Design category C is vessel that operates in seas with significant wave heights up to 2 m and a typical steady wind force of Beaufort Force 6 (12 m/s) or less. While design category D for vessel which operates in seas with significant wave heights up to and including 0.30 m with occasional waves of 0.5 m in height, and a typical steady wind force of Beaufort Force 4 (7 m/s) or less. Since most of fishing vessels in Indonesia are small fishing boat which are less that 5GT, the requirement to provide liferaft on board can be substituted by buoyant apparatus. However, the buoyant apparatus that is provided for safety equipment still should fulfil the standard.

Buoyant apparatus is flotation equipment (other than lifeboats, liferafts, and personal flotation devices) designed to support a specified number of persons in the water, and of such construction that it retains its shape and properties and requires no adjustment or preparation for use.

Based on the aforementioned problem, this study aims to design an FAD that has a floating part so that it can also function as a safety device, in this case a buoyant apparatus. Because of this outcome, the construction of the floating part of FAD must meet the requirements as safety equipment and must have the buoyancy needed to carry FAD equipment.

Table 1. Life-saving appliances for vessels of different design categories

Distance from safe haven	≤ 5 nm	≤ 20 nm	≤ 100 nm	≤ 200 nm	> 200 nm
Liferaft Buoyant apparatus	A⁺, B⁺,	A⁺, B⁺, C♣, D♣	A, B, C, D⁺	A, B, C, D	A, B, C, D
Lifejacket♥	A, B, C♦, D♦♣	A, B, C♦, D♦♣	A, B, C♦, D♦♣	A, B, C, D	A, B, C, D
Immersion suit ♠	A, B	A, B	A, B	A, B	A, B
Lifebuoy •	A, B, C, D	A, B, C, D	A, B, C, D	A, B, C, D	A, B, C, D
Distress signals: 4 parachute rockets⁺⁺⁺	A, B	A, B	A, B	A, B	A, B
Distress signals: 2 handheld flares	C, D	C, D	C, D	C, D	C, D
Capsize rope	A, B, C, D	A, B, C, D	A, B, C, D	A, B, C, D	A, B, C, D
Whistle, mirror, and torch	A, B, C, D	A, B, C, D	A, B, C, D	A, B, C, D	A, B, C, D

Note:
[a] + The liferaft may be substituted with a buoyant apparatus;
[b] ♣ Recommended.
[c] ♥ For every person on board.
[d] ♦ The lifejacket may be substituted with a personal flotation device.
[e] ♠ For every person on board a vessel operating in areas where low water or air temperature can be expected.
[f] Where the vessel is decked and 7 m in LOA or over.
[g] +++ Two of the rockets may be replaced by handheld flares.

The Fish Aggregating Devices (FAD), which was designed in this study, is an FAD which consists of lighting and a source of electricity which was planned in the form of solar panels. Lights are used to attract fish to get closer to the light source, so the fishermen could catch them easily. The method of catching fish using FAD in the form of lights which aims to attract fish to the surface generally uses a certain fishing gear, such as purse seine, or gill net. This method is known to be much more environmentally friendly than the use of Trawl nets which are used to the bottom of the waters and have the potential to damage coral reefs. With the use of fishing gear and fishing aids that are more environmentally friendly, it is hoped that sustainable fishing vessels can be created that support the achievement of a sustainable fisheries sector in Indonesia.

2. Methodology

The design of FAD which also used as buoyant apparatus for safety equipment was divided into two stages. The first stage was designing of floating part, which is at the same time should fulfil the requirement for the buoyant apparatus for safety equipment. The second stage was the arrangement of FAD itself with the lamp and its electrical sources.

Considering that a buoyant apparatus is a safety device that functions as a floating equipment for ship crews in the event of a ship accident, the buoyant apparatus must be designed so that it has the minimum buoyancy required based on the number of crew members, has good stability, and also has good construction strength to withstand the crew's load and the load when it is thrown into the water from the height of the ship.

Furthermore, the design of floating part should consider the minimum standard for safety equipment recommended by International standard, for instance it should meet the standards of SOLAS 74/96 on Life Saving Appliances Code , or the *The Safety Recommendations for Decked Fishing Vessels of Less than 12 metres in Length and Undecked Fishing Vessels* by FAO.

Those requirements including: 1) *It is of such size and strength that it can be thrown from the place where it is stowed into the water without being damaged; 2) it can be stowed where it is readily accessible, can be quickly and easily detached from the vessel and easily launched by hand; 3) it is made of buoyant material and of robust construction; 4) it would be effective and stable when floating either way up*

The design of the FAD system included the calculation of the number of installed lights and the required electrical sources. The source of electricity for FAD was designed coming from solar panels, so at this stage the design of FAD including the calculation of solar panel specifications, the need for batteries and the battery charging system from solar panels.

The weight of the FAD should be considered carefully in which the buoyancy provided by the buoyant apparatus could cover all the complete equipment. Moreover, the buoyant apparatus-FAD module must be tightly bound during operation, thus reducing the possibility of the FAD module being released from the buoyant apparatus and dropped into the sea. Therefore, at this stage the design of the supporting system for the FAD module in the buoyant apparatus and the binding system of the FAD module into the buoyant apparatus structure were carried out as well.

3. Result and Discussions

The buoyant apparatus part was made of solid Polyurethane (PU) and was coated by 2 layer of Fiberglass Reinforced Plastic (FRP). The dimensions of the buoyant apparatus were 1.2 x 0.9 meters with a rectangular cross section measuring 0.25 x 0, 25 meters. It has a line securely attached round the outside, so the crew could hold this line during in the water, as seen in buoyant apparatus design in figure 2(a).

The result from the Maxsurf Software, as seen in table 2, showed that the buoyancy capacity of the designed buoyant apparatus is 138.32 kg. Considering the weight of the FAD module, about 41.5 kg, and the weight of buoyant apparatus itself about 15 kg, then the remain buoyancy available is about 81.82 kg. If considering the buoyance capacity required for one person is about 12.24 – 15,3 kg, then this buoyant apparatus can be used by about 4 – 6 person.

The FAD module consists of 4 @ 10 watt light bulb, 1 unit of 100WP solar panel, and 4 batteries with capacity of 12v 18Ah. The batteries were stored in the box made of FRP. And the lamps were attached on each side of module, as seen in design of FAD Module in figure 2(b). The module was designed can be operated for 10 hours according to the length of fishermen fishing operation in using the FAD for attracting fish.

Buoyant apparatus testing was conducted by throwing the buoyant apparatus into the water from 2 meter height. The purpose of this test is to see the strength of the structure from any damage if it is dropped from the boat to the water. The other tests were related to the ability of the buoyant apparatus to float in the water and its capacity to withstand the load of the crew in the water.

Table 2. Buoyant apparatus hydrostatic characteristic

Measurements	Volume	Units
Displacement	138,32	kg
Volume	0,135	m^3
Draft to Baseline	0,25	m
Lwl	1,2	m
Beam wl	0,9	m
WSA	4,17	m^2
Maximum cross section area	0,112	m^2
Waterplane Area	0,54	m^2
KB	0,125	M
BMt	0,472	m
Immersion (TPC)	0,006	tonne/cm

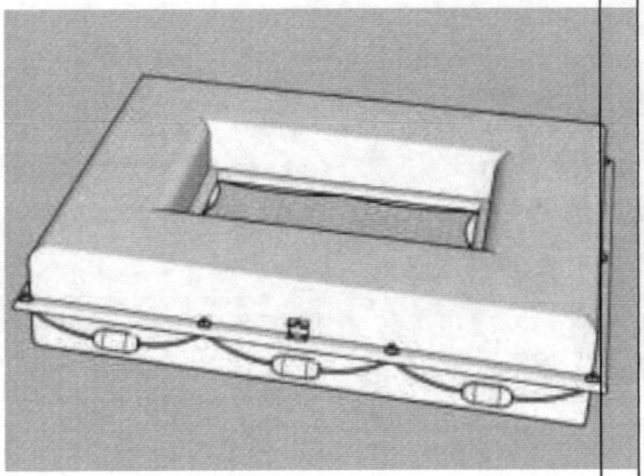

Figure 2. Design of FAD with Buoyant Apparatus (a) Buoyant Apparatus Design

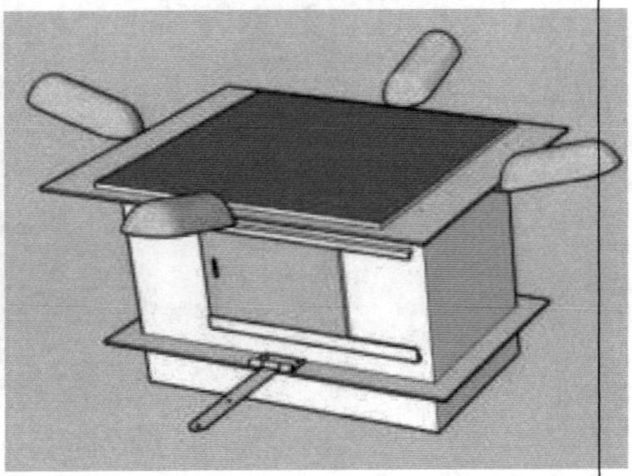

Figure 2. Design of FAD with Buoyant Apparatus (b) Design FAD Module

The drop test showed there was not any damage found on the buoyant apparatus, and therefore it is fulfil the requirement for the structure strength. The floating capacity test was carried out by certain number of person hanging on the line of buoyant apparatus. The result showed that with 4 people hanging on each side of the buoyant apparatus, the depth of the draught of the buoyant apparatus was only less than 5 cm, as seen in figure 3.

Floating test of the complete buoyant apparatus that was attached with the FAD module also carried out with 4 people hanging on each side of the floating structure. For this condition, the draught of the buoyant apparatus was only 8 cm, or one third of the total height, which is 25 cm. It is concluded that the designed buoyant apparatus can be used for more crew, however, by considering the space available, 6 person is maximum capacity for this buoyant apparatus.

To ensure the electrical system of the FAD function well, the test on the light and solar panel system were carried out as well. The test result showed that the system was work properly, as seen in figure 4.

4. Conclusions

The lack of traditional fishermen' awareness to provide safety equipment on their fishing boat has resulted in high rate of fatal accident in fisheries sector in Indonesia. The FAO/ILO/IMO recommendation regarding safety equipment on board allows the use of substitution for certain safety equipment.

Figure 3. Buoyant apparatus test

Figure 4. FAD Lighting system test

The design of buoyant apparatus that also used as floating part of Fish Aggregating Devices is believed can become a solution for low of traditional fishermen' willingness to provide liferaft or buoyant apparatus on their fishing boats. 1.2m x 0.9m size of buoyant apparatus has been constructed to support FAD that consist of 1 unit 100WP solar panel to support 4 light @10watt for 10 hours operation. The floating part can support maximum 6 crew in the water, and easily lift and throw to the water by only one person.

5. References

[1] Suwardjo, D, Haluan, J, Jaya, I and Poernomo, S.a.H 2010 Fishing vessel safety from national and international regulations point of view *Jurnal Teknologi Perikanan dan Kelautan*. **1** (1) pp 1-13

[2] Grahadyarini, B L 2015 Develop self-sufficient fishers Available at: http://www.kiara.or.id/memandirikan-nelayan/

[3] Wibawa, I P A 2016 Sustainable fishing vessel development by prioritising stakeholder engagement in Indonesian small-scale fisheries (PhD Thesis, Newcastle University, UK)

[4] Wibawa, I P A and Birmingham, R W 2018 Improving safety working environment on Indonesian fishing fleet: A case study on local fishing communities in East Java. The 1st Maritime Safety International Confrence *MASTIC 2018* (Bali, Indonesia)

[5] FAO/ILO/IMO 2012 Safety recommendations for decked fishing vessels of less than 12 metres in length and undecked fishing vessels (FAO, Rome)

Acknowledgements
This work was supported by Politeknik Perkapalan Negeri Surabaya, under DIPA Research Funding 2018. Further research is planned for implementation of the product.

Design automatic waitress in android based restaurant using MQTT communication protocol

V Y P Ardhana[1], Tibyani[2], F A S Harianto[3], Sarwadi[4], R A Pratama[5] , I Sutrisno[6], J Endrasmono[7] and R Soekarta[8]

[1,4,5]Universitas Qamarul Huda Badaruddin, Indonesia
[2]Universitas Brawijaya, Indonesia
[3,6,7]Politeknik Perkapalan Negeri Surabaya, Indonesia
[8]Universitas Muhammadiyah Sorong, Indonesia

E-mail: valianypa81@gmail.com

Abstract. Currently, many technologies are created to facilitate and help solve problems in society. The development of technology and industrial automation encourages people to fulfill their daily needs quickly, precisely and efficiently. At this time, a variety of automatic equipment that supports human performance in carrying out their duties has been widely used, ranging from simple control tools to robotics with a variety of up-to-date technology bases. how to queue in front of the cashier and be delivered by the waiter to the dining table which utilizes a lot of Human Resources with various risks such as dishes or even fallen humans. Automatic serving is an alternative for serving food in a modern style. In the process of serving food, it is carried out by placing the food on the robot, using a stepper motor, the robot can walk and by using the photodioda sensor to detect food running on the food tray it will automatically stop in front of the table of customers who have ordered through the android application that has been provided in each table to seat. This system is expected to be able to make the system at the restaurant and its visitors safer and more comfortable.

1. Introduction

The word robot comes from the Czech language, namely robota, which means worker. According to the meaning of language, a robot is a mechanical device that can perform physical tasks, either using human supervision and control, or using predefined programs. Meanwhile, according to the Big Indonesian Dictionary, robots are tools in the form of people and so on that can move or act like humans, and are controlled by machines [1].

Serving food / culinary at this time by queuing in front of the cashier and delivered by the waiter to the dining table. In addition, the waiter brings many dishes regardless of the risk of the dish falling or anything else [2]. The development of increasingly modern technology has made it easier for people today, not only in industry, but in the field of food service as well. This has led to a shift in consumption patterns, people are more interested in eating outside the home in search of something practical. Therefore, to attract consumers, many restaurants try to be superior in providing services. This of course can be overcome by creating an innovation that makes it easier for the community, by utilizing Android and Web applications using the automatic waiter MQTT communication protocol with electronic payment methods. It is hoped that today's restaurant services will be more effective.

Content from this work may be used under the terms of the Creative Commons Attribution 3.0 licence. Any further distribution of this work must maintain attribution to the author(s) and the title of the work, journal citation and DOI.
Published under licence by IOP Publishing Ltd

2. Methodology

In 2018, Food Serving System Prototype Using RFID-Based Arduino discusses creating an automation system for serving food in restaurants using RFID-based arduino. The method used is a laboratory experimental method, namely by creating and conducting a test program to then apply it to the prototype [3].

In 2018, Food Sort Control System Using Android as Human Machine Interface (HMI) discusses the use of android devices as a display interface or commonly known as the Human Machine Interface (HMI). Created with the aim of helping the work of an industrial unit in controlling and monitoring food sorting equipment according to predetermined usage needs [4]. Based on previous research, the authors developed a prototype to implement waiters automatically. This research develops what can be accessed by android applications.

3. Result and Discussion

3.1.System Design

At this stage, the system design is carried out, including mechanical design and electrical design. In addition to making mechanical and electrical designs, making software designs is also needed so that later it can simplify integration between hardware and software.

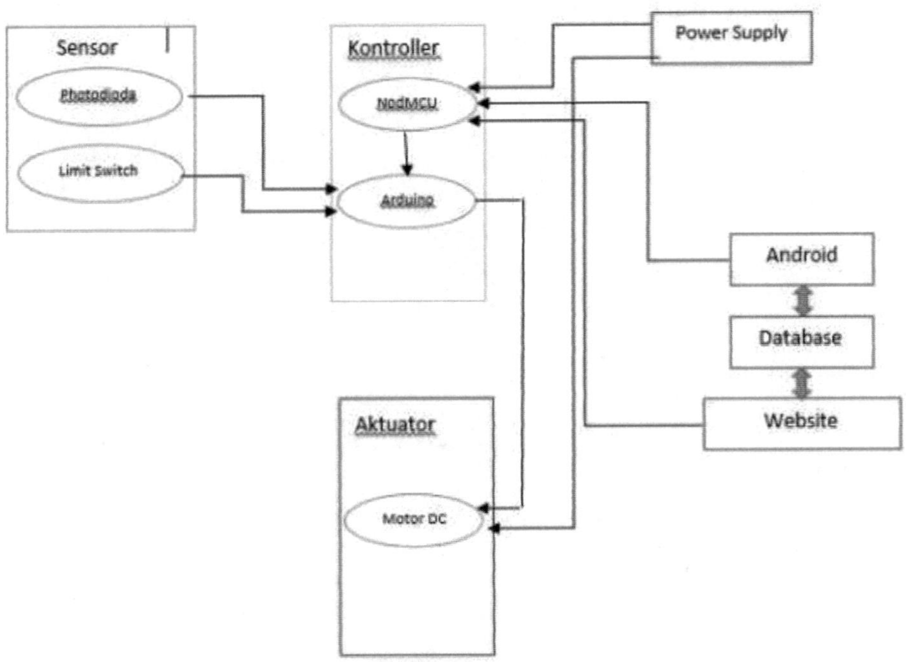

Figure 1. System design

3.2.Prototype Design

In this automatic waiter system using 1 input which will be processed and produce 1 output. The input is in the form of order menu data which is then processed by the smartphone. The smartphone will send the data that will be received by the server computer. Furthermore, the restaurant chef will follow up. After the cooking is ready, the chef immediately places the food and drinks according to the data on the server on the robot. This system will automatically distribute food and beverages in the restaurant according to the table number ordered. Figure 3.2 shows the prototype design that will be used.

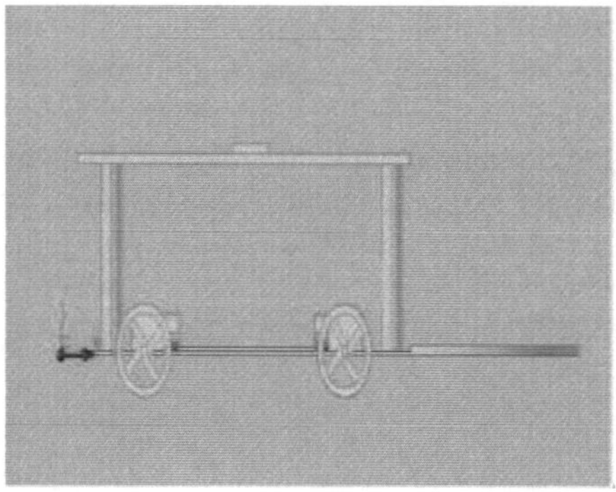

Figure 2. Side view of hardware design

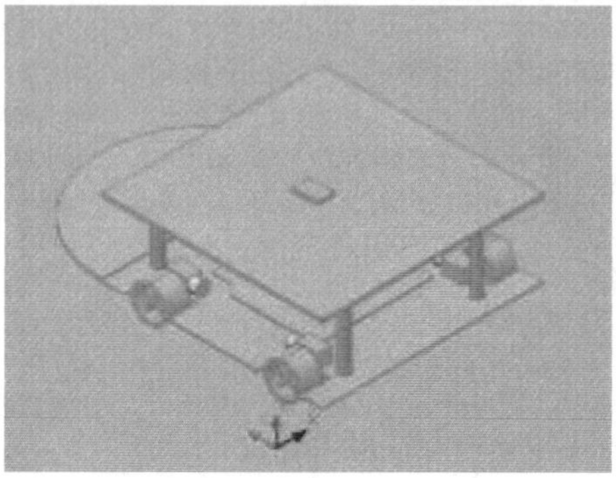

Figure 3. Top view of hardware design

The design of the prototype automatic waiter system consists of 4 wheels for each wheel, there is a 12 V motor for serving food and drinks, delivering according to orders that have been processed by a computer server or cashier with a path as in Figure 3.4.

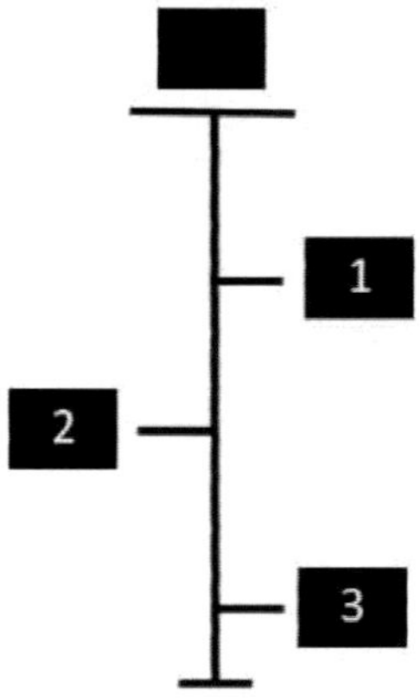

Figure 3. Track path of automatic waiter robot

Figure 3.4 shows the black line as the trajectory of the robot and numbers as the restaurant table number.

3.3. MQTT Protocol Communication

Message Queue Telemetry Transport or commonly called MQTT is a protocol for machine to machine or M2M communication and works at the seventh layer or application and is lightweight message. Even if the connection is lost, all messages sent will be guaranteed by the MQTT protocol. The publish / subscribe communication method is the sending method used by the MQTT protocol. The message on MQTT is sent to the broker and contains the topic sent by the publisher. Then the topic was processed to be forwarded to the subscriber based on user requests.

3.4. Android Application Design

The android application that will be shows from the android application that will be used for the process of ordering food and drinks that have been installed on each device on the table.

Figure 4. Android apllication loading page

Figure 5. Android menu register and login profile

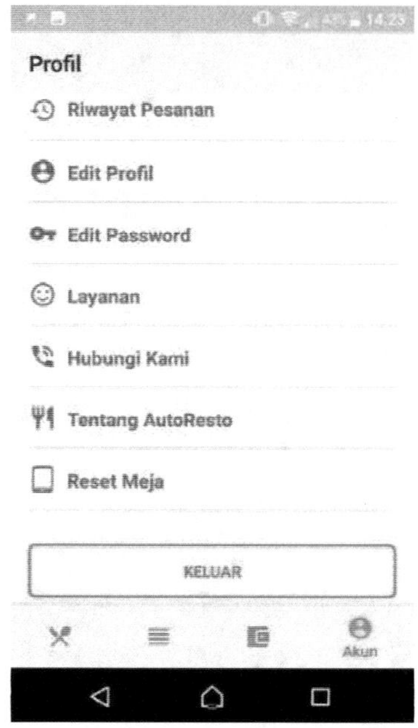

Figure 6. Android application menu user profile

Figure 7. Android application saldo menu

IOP Conf. Series: Materials Science and Engineering 1175 (2021) 012009 doi:10.1088/1757-899X/1175/1/012009

Figure 8. Android order history

To integration communication between the two uses the MQTT communication protocol where the application will publish messages on predetermined topics into the broker. Then the broker will receive the message and send it to nodemcu to command the prototype. The order data will be received by the chef and stored automatically into the database shown in Figure 3.9.

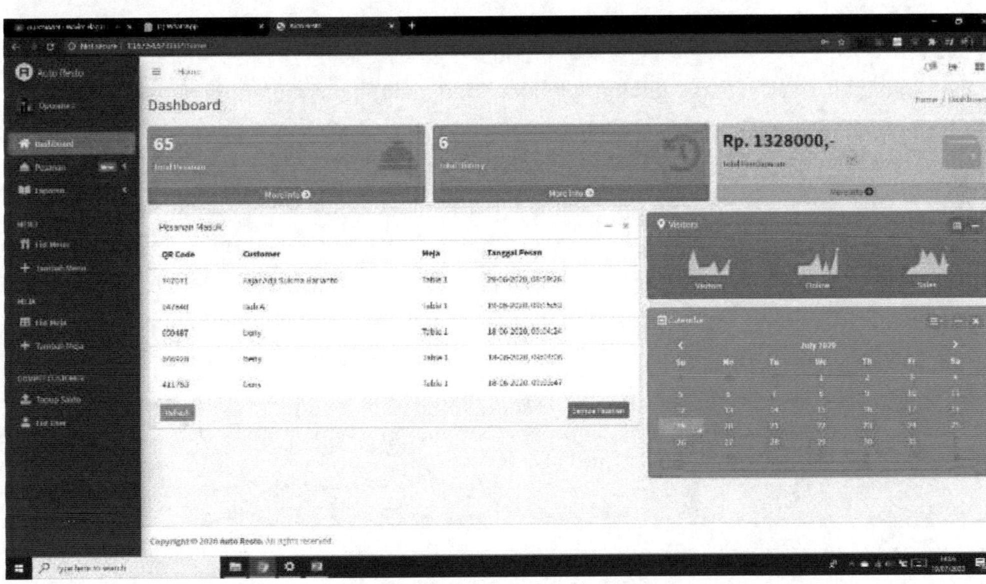

Figure 9. Main of website

The website created as shown in the image above will be used by the cashier and to process orders according to the data received from the customer to the cashier.

IOP Conf. Series: Materials Science and Engineering 1175 (2021) 012009 doi:10.1088/1757-899X/1175/1/012009

3.5. Trial and Analysis

In the implementation stage the system has been adjusted to the restaurant scheme. There are 3 tables as places for distributing orders. Arduino software is used as a programming language to retrieve data in the monitoring process. The prototypes that have been made and implemented in this system. The next stage after the prototype and application are ready for use is to test and analyze the system so that it runs well. Various kinds of tests are carried out, including testing the application function, namely testing the error magnitude in prototype integration with the system that may occur and can run according to the flowchart that has been compiled. In the system integration process, the application will publish messages according to predetermined topics as shown in Figure 10.

Figure 10. Publish subscribe system

In Figure 10. above we can see that predetermined topics can be published and subscribed by the system, so that when the application places an order, the order is automatically received by the restaurant and then distributed according to the order. In Figure 9. it can be seen that predetermined topics can be published and subscribed by the system, so that when the user application places an order at the restaurant table, the order will automatically be received by the restaurant as shown in Figure 11. and Figure 12.

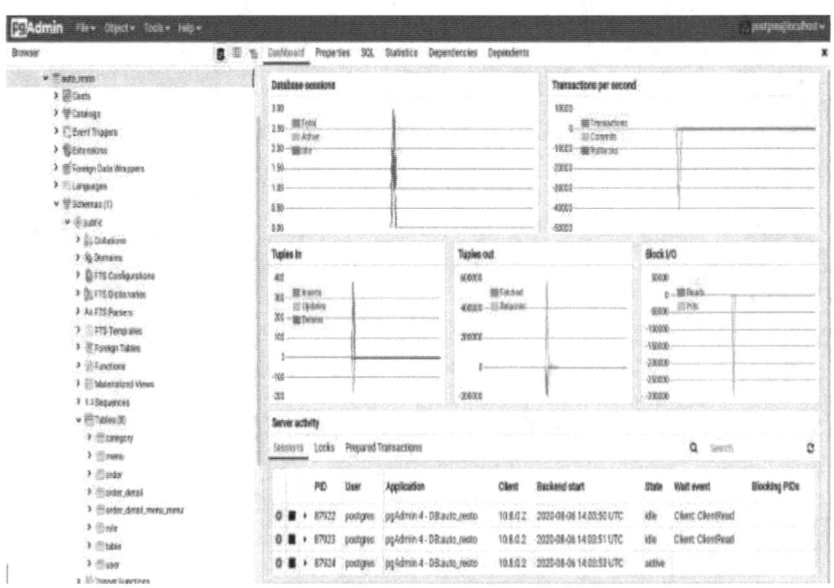

Figure 11. Table building

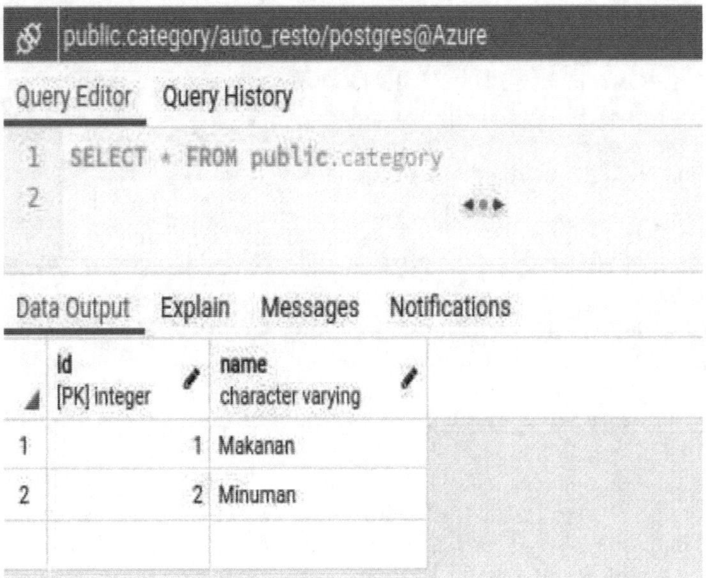

Figure 12. Database

Figure 11. and Figure 12. is the result of data in the database which shows the order information made by the application and will then be received by the system. The Autoresto application has been tested to record whether the application can run according to the flowchart that has been compiled and produces the output value according to the user's wishes.

Furthermore, testing is also carried out on the website to find out whether the website is able to accept orders according to the orders that have been sent from the application, so that the chef can serve the appropriate orders and distribute them to the restaurant table. In the last stage of testing the communication protocol used, to determine the delay and error generated by the MQTT communication protocol.

4. Conclusion

The MQTT protocol used is able to run well and smoothly when implemented in the IoT device monitoring system. The publish-subscribe principle that is applied can work well in the data transfer process that has been done and obtained a topic that contains information needed by the user based on observations from the specified device. Also concluded that the resources in the MQTT protocol can work well in low bandwidth conditions and high latency.

5. References

[1] Mesin T, Teknik F and Malang U M 2017 *Rancang Bangun Belt Conveyor Untuk Penyaji.* p 1 [Accessed: 26-Jan-2020]

[2] Danis D A N, Pang D, St J O W and Elektro J T 2014 Rancang bangun robot pengantar makanan line follower. *E-Journal Teknik Elektro dan Komputer.* **3** 1 pp 1-10 [Accessed: 09-Jan-2020]

[3] Mesin T, Teknik F and Malang U M 2017 *Rancang Bangun Belt Conveyor Untuk Penyaji.* pp 1-12 [Accessed: 26-Jan-2019]

[4] Jami'in M A, Hu J, Marhaban M H, Sutrisno I and Mariun N B 2016 Quasi ARX neural network based adaptive predictive control for nonlinear systems. *IEEJ Trans Elec Electron Eng.* **11** pp 83-90

[5] Jami'in M A, Sutrisno I and Hu J 2012 Lyapunov learning algorithm for Quasi-ARX neural

network to identification of nonlinear dynamical system. *2012 IEEE International Conference on Systems, Man, and Cybernetics (SMC)* (Seoul, 2012) pp 3147-3152 doi: 10.1109/ICSMC.2012.6378275

[6] Jami'in M A, Sutrisno I and Hu J 2013 Deep searching for parameter estimation of the linear time invariant (LTI) system by using Quasi-ARX neural network. *The 2013 International Joint Conference on Neural Networks (IJCNN)* (Dallas TX, 2013) pp 1-5 doi: 10.1109/IJCNN.2013.6707091

[7] Jami'in M A, Sutrisno I and Hu J 2015 Maximum power tracking control for a wind energy conversion system based on a quasi ARX neural network model. *IEEJ Trans Elec Electron Eng.* **10** pp 368-375

[8] Jami'in M A, Sutrisno I, Hu J, Mariun N B and Marhaban M H 2014 An adaptive predictive control based on a quasi-ARX neural network model. *13th International Conference on Control Automation Robotics & Vision (ICARCV)* (Singapore, 2014) pp 253-258 doi: 10.1109/ICARCV.2014.7064314

[9] Sutrisno I, Jami'in M A and Hu J 2014 Modified fuzzy adaptive controller applied to nonlinear systems modeled under quasi-ARX neural network. *Artif Life Robotics* **19** pp 22–26. https://doi.org/10.1007/s10015-013-0137-6

[10] Sutrisno I, Jami'in M A and Hu J 2014 An improved elman neural network controller based on quasi ARX neural network for nonlinear systems. *IEEJ Trans Elec Electron Eng.* **9** pp 494-501

[11] Sutrisno I, Che C and Hu J 2014 An improved adaptive switching control based on quasi ARX neural network for nonlinear systems. *Artif Life Robotics.* **19** pp 347–353

[12] Sutrisno I, Jami'in M A and Hu J 2012 Neural predictive controller of nonlinear systems based on quasi-ARX neural network. *18th International Conference on Automation and Computing (ICAC)* (Loughborough, 2012) pp 1-6

[13] Rinanto and Khumaidi *Aplikasi Android- Rasspberry Pi pada Kapal Tanpa Awak Untuk Pencarian Korban Kecelakaan Laut*

[14] Widodo R B, Quita R M, Amrizal S, Gunawan R S, Wada C, Ardiansyah A, Sutrisno I, Rahmat M B, Arfianto A Z, Handoko C R, Santoso A D, Santosa A W B, Ardhana V Y P and Setiawan E Grasping and attached mode in human computer interaction in the study of mouse substitution. *Journal of Physics: Conference Series* **1196** 1 012034

[15] Sutrisno I, Jami'in M A, Hu J and Marhaban M H 2015 Self-organizing quasi-linear ARX RBFN modeling for identification and control of nonlinear systems. *54th Annual Conference of the Society of Instrument and Control Engineers of Japan (SICE)*. IEEE

[16] Jami'in M A, Sutrisno I and Hu J 2014 Nonlinear adaptive control for wind energy conversion systems based on quasi-ARX neural networks model. *Proceedings of the International Multi Conference of Engineers and Computer Scientists.* (March 12-14 2014, Hong Kong)

[17] Sutrisno, I 2009 Pemrograman Komputer dengan Software Matlab (Surabaya: ITS Press)

IOP Conf. Series: Materials Science and Engineering 1175 (2021) 012010 doi:10.1088/1757-899X/1175/1/012010

Safe distance reminder system on ship against port for the standing process using image processing

A W B Santosa[1], A Hardianti[2], S Hasugian[3], I Sutrisno[4] and A Khumaidi[5]
[1] Universitas Diponegoro, Indonesia
[2,4,5]Politeknik Perkapalan Negeri Surabaya, Indonesia
[3] Politeknik Pelayaran Surabaya, Indonesia

E-mail: Arikapal75@gmail.com

Abstract. Since ancient times, Indonesia has been known as a country that conducts trading activities. However, the number of accidents when the ship is about to dock is an indication of the need to improve the marine transportation system. The factors that cause accidents include human, technical and weather errors. Seeing these conditions and also related to the increasingly rapid development of technology, modern solutions for detecting objects using cameras can be developed. By using a PC as a video processor from a camera that works in real time and can classify several ship objects using the Convolutional Neural Network (CNN) method. As well as being able to estimate the distance of the object where the ship must turn off the engine before docked with the camera using Stereo Vision. and the results of the algorithm processing can be translated into a signal that will be sent to the harbormaster and also the ship that will dock as an output. Then the ship that will dock can find out the distance where the ship has to turn off the engine through the signal sent or siren. The average total accuracy of the system which is able to detect ships is 97.15%.

1. Introduction

Indonesia is the biggest archipelago country in the world. Since ancient times, Indonesia has been known as a country that has carried out many trade activities with water routes. Along with the times, now the world of world shipping is experiencing rapid development. Trade using sea transportation services is an option because the costs are cheaper or lower than other transportation services. The Port of Tanjung Perak is one of the gateways for sea transportation for eastern Indonesia, acting as a center for collectors and distributors of goods from Eastern Indonesia, particularly from East Java Province.

One of the facilities at the terminal is a pier, where the dock is a place to dock and unload ships. The jetty at the terminals at Tanjung Perak Port in Surabaya has different sizes. If the dock available at a terminal is longer than the other terminal, then the terminal can accommodate more ships than other terminals. The depth must not be less than 1.15 times the maximum draft of the largest vessel and must not be more than 100 m (Graillot, A., 1983). But the size of the ship also affects the number of ships that will moor at the dock. If ships with large sizes moor at the terminal, the terminal can only accommodate a small number of ships. So there are two important things that become the benchmark for ships that will moor, namely the length of the pier and the size of the ship. If the ship to be moored does not have a dock that matches the size of the ship, then the ship must wait. This will result in queues for ships afterwards and can also be the cause of ships colliding.

Content from this work may be used under the terms of the Creative Commons Attribution 3.0 licence. Any further distribution of this work must maintain attribution to the author(s) and the title of the work, journal citation and DOI.
Published under licence by IOP Publishing Ltd

With the conditions and technological developments as above, the authors have the idea to create a tool to detect objects that will enter the dock using video processing which is supported using Stereo Vision. This prototype uses a camera to detect moving objects and monitor the boundaries of the area. Personal Computer (PC) is used to process images captured by cameras that can detect and classify a ship or moving object using the Convolutional Neural Network (CNN) method. So that the prototype can maximize the performance of the harbormaster in order to facilitate the work of the process of getting in and out of the ship at the dock can be optimal and minimize the occurrence of accidents.

2. Methodology

2.1. System Design

Convolutional Neural network is an architectural extension of feed forward artificial neural network with a larger number of layers. The CNN architecture is determined by the number of layers, the number of maps in each layer, the size of the kernel, the size of the subsampling windows, and how they are connected to each other. In Figure 2.7. You can see how the Convolutional Neural network architecture works in recognizing handwriting.

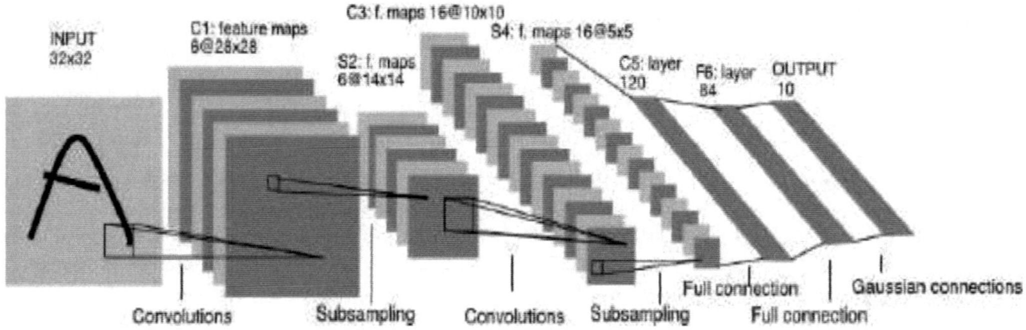

Figure 1. Arsitektur LeNet5

2.1.1. System Design.

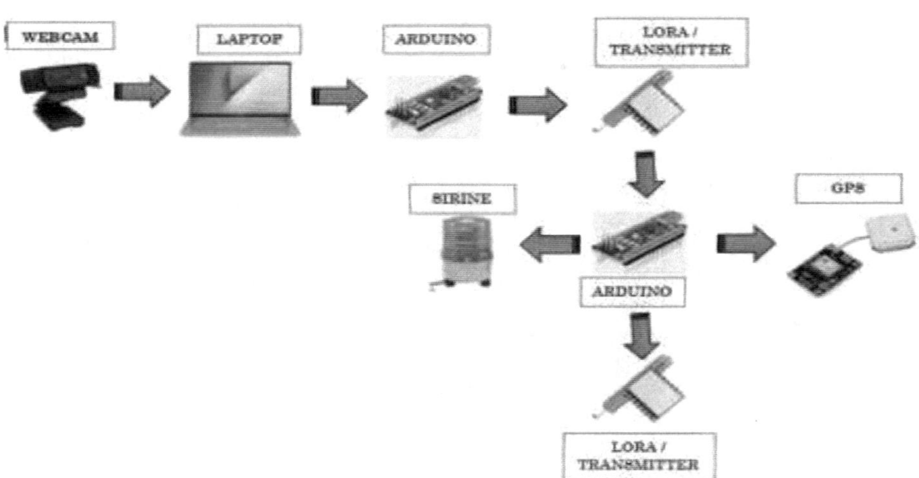

Figure 2. System Design

Nama Dataset	Gambar Dataset	Jumlah Dataset
SHIP	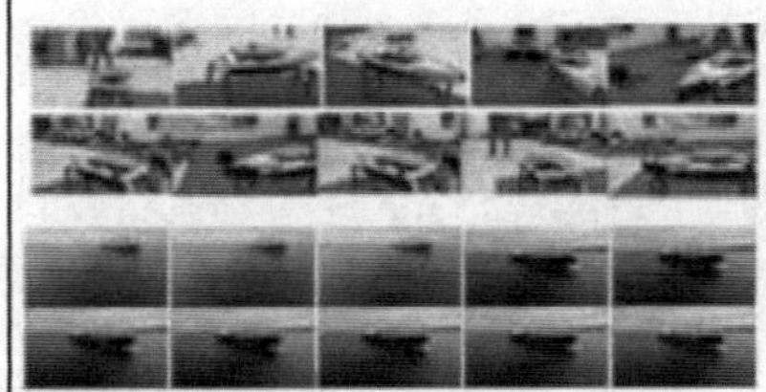	500
NOT SHIP		500

Figure 3. Dataset

Dataset is an object that represents data and relations that have been created in storage. A dataset can be called a database, because it has the same structure. The dataset in this study is taken from several examples of several images that have been stored in the laptop storage space. And the dataset is used for object classification. the more input the dataset will make the higher the classification results and the heavier the laptop runs.

2.1.2. CNN Architechture.

This final project is focused on getting an object detection system with a camera. In addition, the main objective of this study is to obtain a classification result from the CNN process. In the CNN process in general, this process has 3 stages, namely pre-processing, processing, and classifying.

2.1.3. Euclidean Distance Method.

Euclidean Distance method is a method related to distance calculation. This concept was introduced by a mathematician named Euclid of Alexandria around 300 B.C.E (Wikipedia, n.d.-c). The principle of the Euclidean method is to find the proximity of the distance value between two variables. The following is the formula for measuring the distance of Euclidean Distance.

IOP Conf. Series: Materials Science and Engineering 1175 (2021) 012010 doi:10.1088/1757-899X/1175/1/012010

$= \sqrt{} (Lat_1 - Lat_2)\ 2 + (Long_1 - Long_2)^2$

The formula above is a formula used to find distance values based on two coordinate points based on sources (Adiwilaga, 2014). Based on this formula, the data from the known ship distance values will then be compared to get the closest value to the ship.

3. Result and Discussion

3.1. GPS Testing Result

GPS testing, using ublox neo M8N GPS and Mobile GPS to find out the error value or difference obtained. Here are the points taken from the neo M8N GPS and Mobile GPS. The data used in this research is data obtained directly from the GPS device. The microcontroller will receive laTitude and longitude data in the form of a decimal degree. From the latitude and longitude data will be processed into the length of the distance in kilometers. Where the authors here process the data using the Euclidean Distance Method.

3.1.1. Euclidean Method

$$s = (\sqrt{} ((x2\text{-}x1) \wedge 2 + (y2\text{-}y1) \wedge 2)) * r \quad (1)$$

Where :

s = distance (km)	$y1$ = longitude 1
$x2$ = latitude 2	$y2$ = longitude 2
$x1$ = latitude 1	r = 1° earth = 111,319 km

Table 1. Neo M8N GPS test points

Keterangan	**Latitude**	**Longitude**	**Hasil Pengujian Jarak Metode Ecluidean**
Titik 1	-7.788983	112.009176	1.168,84 Cm
Titik 2	-7.789044	112.009132	

Table 2. Test points for mobile GPS

Keterangan	Latitude	Longitude	Hasil Pengujian Jarak Metode Ecluidean
Titik 1	-7.788950	112.009119	510,13 Cm
Titik 2	-7.788964	112.009151	

From the test table above, the error difference between GPS Neo M8N and GPS Hanphone is obtained, namely:

= 1168.84 Cm - 510.13 Cm

= 698.72 cm

3.1.2. MobileNet Extracion Features

MobileNet SSD consists of 13 layer blocks which contain a depthwise separable convolution layer.

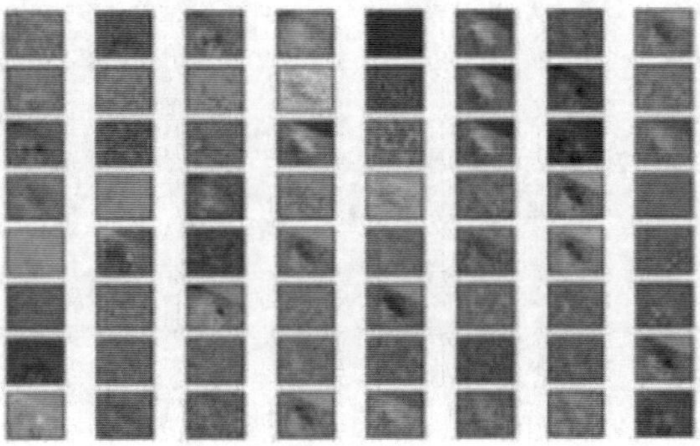

Figure 4. Layer blocks which contain a depthwise separable convolution layer

The image input becomes the first layer, then the depthwise convolution layer will process the image based on the RGB color channel (Red, Green, Blue). Figures below show the map feature of the activation function (RELU6) of the first three layers of the MobileNet architecture used for Ship detection. The map feature is a feature that has been successfully extracted from each layer. From this figure, there are 64 features extracted at each activation layer according to the output size on that layer, which is 64, which means 8x8.

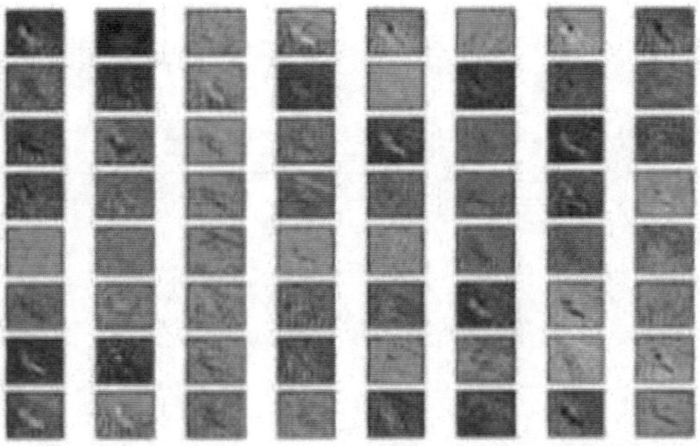

Figure 5. Layer 1 Extraction Feature Function

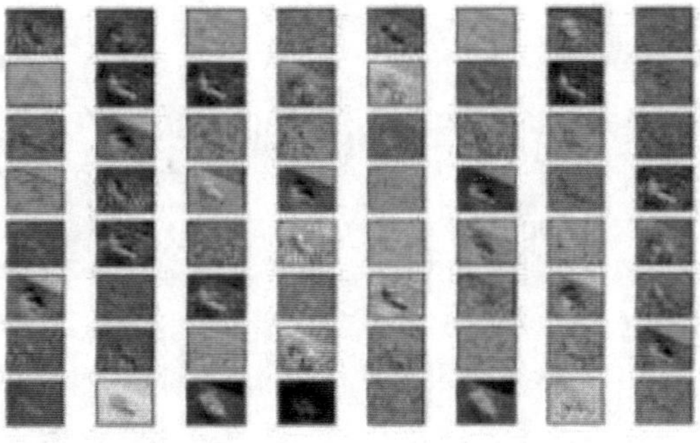

Figure 6. Layer 2 Extraction Feature Functions

IOP Conf. Series: Materials Science and Engineering 1175 (2021) 012010 doi:10.1088/1757-899X/1175/1/012010

Figure 7. Layer 3 Extraction Feature Functions

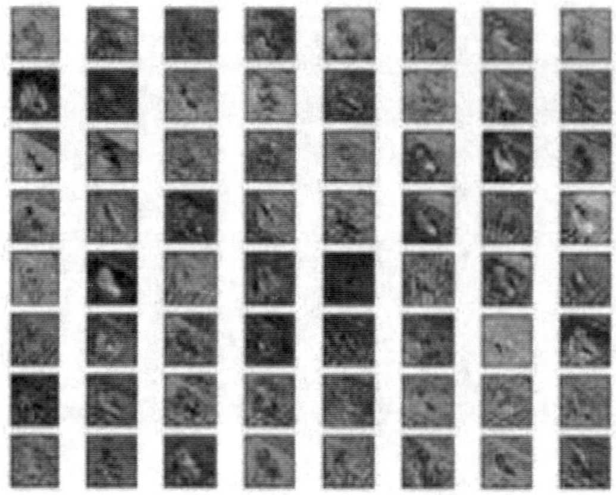

3.1.3. Evaluation Training

The training process is a process in which the processor learns the features of the input data so that the resulting system can recognize the specified object. In this study, the training process was carried out up to 40,000 steps and the network used was Mobilenet SSD. To be more optimal, the training process using a PC with an NVIDIA GeForce GTX 950 M GPU and the total time spent completing 40,000 steps is 8 hours.

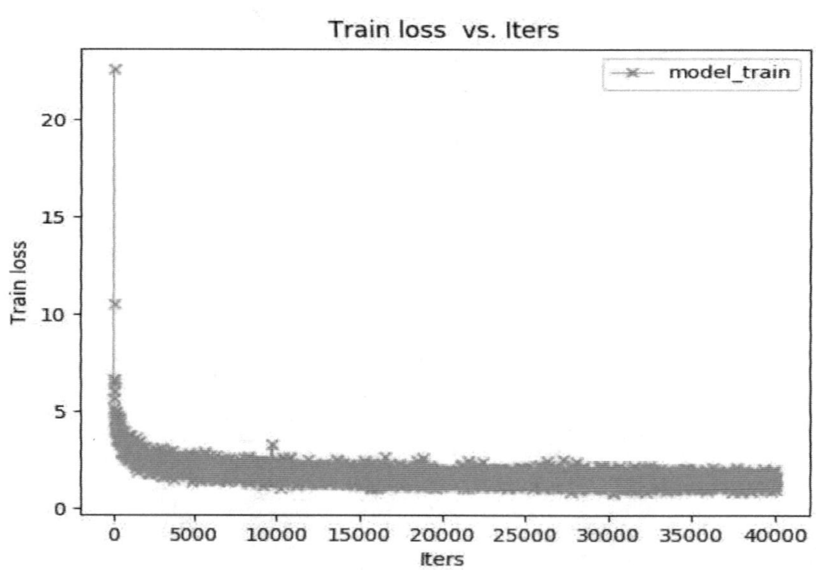

Figure 8. Chart of training loss against iteration

As previously explained, In this architecture, the author uses a multistep learning rate type with a step in the 20,000 and 40,000 iterations. In the 0-20,000 iteration, the initial learning rate used is 0.0005. Then in the iteration of 20,000 - 40,000 the learning rate value changes to 0.00025 because it is multiplied by the value γ.

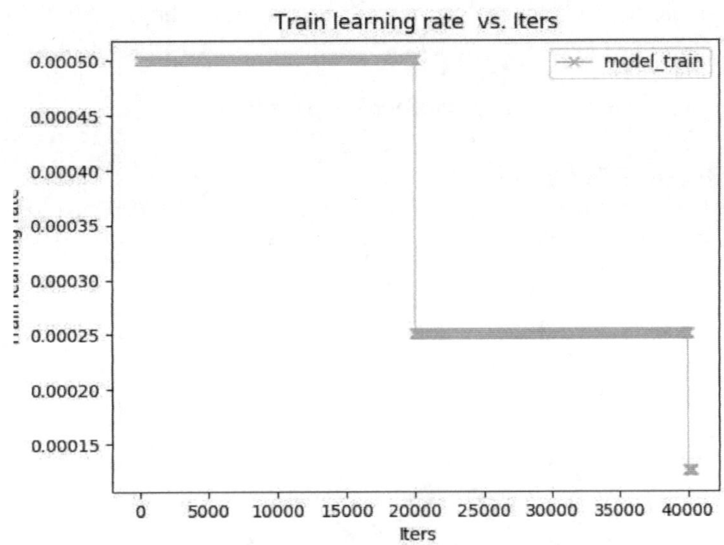

Figure 11. Changes in learning rate value against iteration

The initial learning rate value cannot be too small, because it will make the learning model not comprehensive in the initial phases. It is better if the learning rate value also decreases with the increase in iterations. Figure 11. shows the change in the value of the learning rate with respect to iterations. By using a configuration matrix, precision, recall, and accuracy values can be obtained.

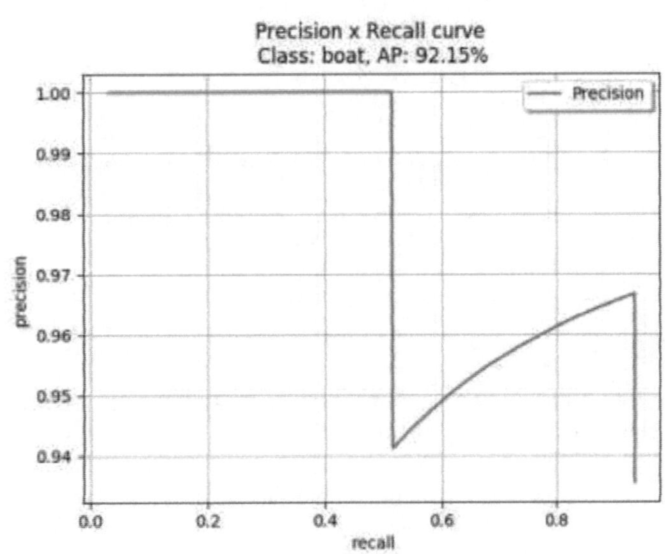

Figure 12. Precision-Recall graph of model validation

In this project, the ship detection model is evaluated to obtain precision and recall values. In finding the precision and recall values, there are 4 parameters that must be fulfilled, namely true positive, true negative, false positive, and false negative. In this case, true positive is the prediction indicating Ship and groundtruth also indicating Ship. True negative does not predict Ship and groundtruth also does not indicate Ship. False positives are Ship's predictions but groundtruth doesn't show Ship. And false negative is not predicting the ship, but the groundtruth shows the ship.

The precision and recall values are calculated for all test data, which is about 530 images. From 530 images, it will produce mAP (Mean Average Precision), which is the average precision value of all test data. The mAP value on this ship detection model is 92.15%.

3.1.4. Vessel Detection Test Results

Ship detection testing is carried out to assess the system's ability to recognize objects in the form of ships. In this test, the confidence value will be obtained, which is the percentage of the model's prediction on the object of the ship.

Table 3. Graph of precision-Recall of model validation

Input Gambar	Hasil Deteksi	*Confidence*
		99,42%
		99,86%
		100%

From the table above, it can be seen that the system's ability to detect ships is quite good, with an average confidence of above 90%.

3.1.5. Ships Detection Test Results from Various Distances

The second test is to vary the distance between the object and the camera when taking pictures. The variations in the distance used are 1 meter, 2 meters, 3 meters, 4 meters and 5 meters. This is done to assess that the system can detect ships from a maximum distance of 5 meters. The test results of ship detection against distance can be seen in table 4 below.

Table 4. Ships detection test results from various distances

Input	Distance	Confidence	Evidence
	1 meter	99,42%	Detected
	2 meter	100%	Detected
	3 meter	99,86%	Detected
	4 meter	51,29%	Detected
	5 meter	25,13%	Detected

From the table we can see that the system's optimal ability to detect ships is at a distance of 1 to 3 meters. This is evidenced by the confidence value that reaches more than 90%. At a distance of 2 meters, the system's ability to detect ships began to decline as evidenced by the decrease in the system's percentage of detecting ships. This can be due to the lack of variation in training data from various distances.

3.1.6. Testing On Real Time Video

After testing the system integration from the input in the form of an image. The next test is to calculate the fps (frames per second) from the input in the form of real time video. The fps test results from real time video can be seen in table 5.

From the table 5, the test results are obtained through real time video. From a total of 5 tests in the table, the system can recognize 3 ships with an accuracy above 90% and 2 ships with an accuracy below 50%. The focus of this test is to prove the system can work in real time as shown by fps. The system can work in real time with a minimum fps value of 4.77 and the highest fps value is 11.90.

Table 5.Testing Results on Real Time Video

Input	Confidence	FPS	Evidence
	96.51%	4.77	Detected
	100%	6.53	Detected
	91.37%	11.90	Detected
	49.5%	11.47	Detected
	46.89%	11.88	Detected

From all test data starting from ship detection, the overall accuracy value of the system is 78.9% and the system speed is 9.31 fps.

4. Conclusion

1. Ship recognition systems can be built with Mobilenet SSDs for detection on Ships. This method was chosen because it can recognize objects well, and is light and fast computational.

2. The results of the matrix evaluation for the ship detection model in the form of mAP (Mean Average Precision) of 92.15% were tested on 530 images.
3. The overall evaluation result of the ship recognition system is 78.9%, where the system is able to recognize ships with a level of truth of 78 per 100 characters.
4. The system can be implemented in real time and the computation rate in one frame is 9.31 frames per second.
5. The optimal distance between the camera and the object is under 5 meters.

5. References

[1] Khumaidi A 2017 Welding defect classification based on convolution neural network (CNN) and Gaussian Kernel. *International Seminar on Intelligent Technology and Its Application.* pp 261-265

[2] Deshpande A 2016 *A Beginner's guide to understanding convolutional neural networks Part 2.* Available at: https://adeshpande3.github.io/A-Beginner%27s-Guide-To-Understanding-Convolutional-Neural-Networks- Part-2/

[3] Khumaidi A 2015 *Implementasi pengolahan video dengan metode color threshold pada prototype kapal pendeteksi korban kecelakaan laut berbasis android.* (Surabaya: Politeknik Perkapalan Negeri Surabaya)

[4] Lutfi M N 2018 *Pendeteksi objek api dan boneka pada robot pemadam api berkaki menggunakan video processing berbasis mini PC.* (Surabaya: Politeknik Perkapalan Negeri Surabaya)

[5] Marzuqi I, Arinata G P, Putra Z and Khumaidi A 2017 Segmentasi dan estimasi jarak bola dengan robot menggunakan stereo vision. *5th Indonesian Symposium on Robotic Systems and Control.* pp 140-144

[6] Salsabila 2018 *Penerapan deep learning menggunakan convolutional neural network untuk klasifikasi citra wayang Punakawan.* (Yogyakarta: Universitas Islam Indonesia)

[7] R Girshick J, Donahue T Darrell and J. Malik 2016 Region-based convolutional networks for accurate object detection and segmentation. *IEEE Trans Pattern Anal. Mach. Intell*

[8] Z Q Zhao, P Zheng, S T Xu and X Wu 2019 Object detection with deep learning: A review. *IEEE Transactions on Neural Networks and Learning Systems*

[9] Jami'in M A, Hu J, Marhaban M H, Sutrisno I and Mariun N B 2016 Quasi ARX neural network based adaptive predictive control for nonlinear systems. *IEEJ Trans Elec Electron Eng.* **11** pp 83-90

[10] Jami'in M A, Sutrisno I and Hu J 2012 Lyapunov learning algorithm for Quasi-ARX neural network to identification of nonlinear dynamical system. *IEEE International Conference on Systems, Man, and Cybernetics (SMC).* (Seoul). pp 3147-3152. doi: 10.1109/ICSMC.2012.6378275

[11] Jami'in M A, Sutrisno I and Hu J 2013 Deep searching for parameter estimation of the linear time invariant (LTI) system by using Quasi-ARX neural network. *The 2013 International Joint Conference on Neural Networks (IJCNN).* (Dallas, TX) pp 1-5. doi: 10.1109/IJCNN.2013.6707091

[12] Jami'in M A, Sutrisno I and Hu J 2015 Maximum power tracking control for a wind energy conversion system based on a quasi ARX neural network model. *IEEJ Trans Elec Electron Eng.* **10** pp 368-375

[13] Jami'in M A, Sutrisno I, Hu J, Mariun N B and Marhaban M H 2014 An adaptive predictive control based on a quasi-ARX neural network model. *13th International Conference on Control Automation Robotics & Vision (ICARCV).* (Singapore). pp 253-258 doi: 10.1109/ICARCV.2014.7064314

[14] Sutrisno I, Jami'in M A and Hu J 2014 Modified fuzzy adaptive controller applied to nonlinear

systems modeled under quasi-ARX neural network. *Artif Life Robotics.* **19** pp 22–26 https://doi.org/10.1007/s10015-013-0137-6

[15] Sutrisno I, Jami'in A M and Hu J 2014 An improved elman neural network controller based on quasi ARX neural network for nonlinear systems. *IEEJ Trans Elec Electron Eng.* **9** pp 494-501

[16] Sutrisno I, Che C and Hu J 2014 An improved adaptive switching control based on quasi-ARX neural network for nonlinear systems. *Artif Life Robotics.* **19** pp 347–353

[17] Sutrisno I, Jami'in M A and Hu J 2012 Neural predictive controller of nonlinear systems based on quasi-ARX neural network. *18th International Conference on Automation and Computing (ICAC).* (Loughborough) pp 1-6

[18] Rinanto N and Khumaidi *Aplikasi Android- Rasspberry Pi pada Kapal Tanpa Awak untuk Pencarian Korban Kecelakaan Laut*

[19] Widodo R B, Quita R M, Amrizal S, Gunawan R S, Wada C, Ardiansyah A, Sutrisno I, Rahmat M B, Arfianto A Z, Handoko C R, Santoso A D, Santosa A W B, Ardhana V Y P and Setiawan E Grasping and attached mode in human computer interaction in the study of mouse substitution. *Journal of Physics: Conference Series* 1196 1 012034

[20] Sutrisno I, Jami'in M A, Hu J and Marhaban M H 2015 Self-organizing quasi-linear ARX RBFN modeling for identification and control of nonlinear systems. *54th Annual Conference of the Society of Instrument and Control Engineers of Japan (SICE).* IEEE

[21] Jami'in M A, Sutrisno I and Hu J 2014 Nonlinear adaptive control for wind energy conversion systems based on quasi-ARX neural networks model. *Procedings of the International Multi Conference of Engineers and Computer Scientists.* March 12-14 2014 (Hong Kong)

[22] Sutrisno I 2009 *Pemrograman komputer dengan software matlab.* (Surabaya: ITS Press)

IOP Conf. Series: Materials Science and Engineering 1175 (2021) 012011 doi:10.1088/1757-899X/1175/1/012011

Design of printed circuit board detection with image information technology

Tibyani[1], N Hayati[2], I Sutrisno[3], A Khumaidi[4], M A Rahman[5], F Gumelar[6], K Dhipanusa[7], Filemon[8] and E Supriyanto[9]
[1]Faculty Of Computer Science, Brawijaya University, Indonesia
[2,3,4]Politeknik Perkapalan Negeri Surabaya, Indonesia
[5,6,7,8]Politeknik Pelayaran Sorong, Indonesia
[9]Politeknik Negeri Semarang, Indonesia

E-mail: tibyani@ub.ac.id

Abstract. In this research, an automatic PCB,s deffects checking tool will be made with image processing technology that adopts Haar Cascade Classifier method as a decision maker for whether or not a broken circuit on the PCB, with this technology the PCB will be detected using a square feature to show the broken path. will be displayed on the interface on the monitor. From the results of the test, data using the Haar Casscade Classifier method has the average percentage of successful detection of broken paths on the PCB at a camera distance of 26 cm and 16 cm is 52.49% with the maximum hassle obtained when the camera distance is placed at a distance of 26 cm and an angle position of 105 degrees which has a success rate of 97.5%. This method is good enough for PCB defect detection tools.

1. Introduction

Today there are still many printed circuit boards produced or in this book it will be abbreviated as PCB conventionally. PCB is a board used to connect electronic components with the layer of the conductor, usually copper. The process of making PCB is conventionally carried out in several basic stages, namely the design stage of the copper path shape and the PCB etching stage. The etching stage is the process of removing unwanted copper from the PCB board. One way to do this etching process is to use ferric chlorite ($FeCl3$) which is dissolved in hot water. During the etching stage, sometimes the etching process doesn't happen well. This can cause defects in the copper lines on the PCB, defects in the copper lines on this PCB can be in the form of broken copper lines or attached copper lines due to undesirable dissolving of the copper layer resulting in short conditions.

When used in the production of electronic equipment, PCBs are very influential in the manufacture of these electronic devices, for example when there are only a few PCB lines that are disconnected or damaged, the electronic device cannot be operated properly. So it is very important in the PCB Quality Check process to check whether the PCB is damaged or not. Usually, in PCB inspection, only direct checking is used in conventional ways.

In large scale PCB production especially for single layer PCB and conventional PCB printing methods. Checking the broken paths on the PCB using conventional methods cannot be done optimally. One of them is due to fatigue, decreased accuracy and lack of resources.[1] From the results of a survey conducted at a company that produces PCBs, it is known that PCB quality control is still done

Content from this work may be used under the terms of the Creative Commons Attribution 3.0 licence. Any further distribution of this work must maintain attribution to the author(s) and the title of the work, journal citation and DOI.
Published under licence by IOP Publishing Ltd

IOP Conf. Series: Materials Science and Engineering 1175 (2021) 012011 doi:10.1088/1757-899X/1175/1/012011

manually and sometimes human errors occur where the error in the percentage value reaches 10%. Therefore, we need a system that can perform PCB path checks quickly and precisely.

Therefore, in this study author tries to make and analyze a PCB defect checker with the help of a camera to replace human vision and to make it easier to apply this PCB checker, a technology called MiniPC or Raspberry Pi is used. With these two technologies, Image Processing can be used to detect objects with the OpenCv and Tensorflow libraries. [2] According to Hanugra Aulia Sidharta, OpenCV (Open Source Computer Vision Library) is an open source library developed by Intel that focuses on simplifying programming related to digital images (Sidharta, 2017). From the use of the OpenCv library, images can be obtained in real time.

With the above technology, the authors will conduct research to make a PCB defect detector with the help of Image Processing with the Haar Cascade method to help determine broken paths on the PCB.

2. Methodology

2.1. System Design

From Figure 1 the systems work first initial in all the system make sure all the hardware connection can work properly. secondly make sure camera censor can work well for cathing image and already installed in raspberry Pi after that the image that already taken by camera will be processed by image processing using OpenCV and all the library that already installed such tensor flow, PIL, and tesseract. PCB's image will be processed by Haar Cascade Classifier Method using the datasets that already train before. If there are defects on the PCB's Circuit as dataset then the deffect's will be shown in the Interface

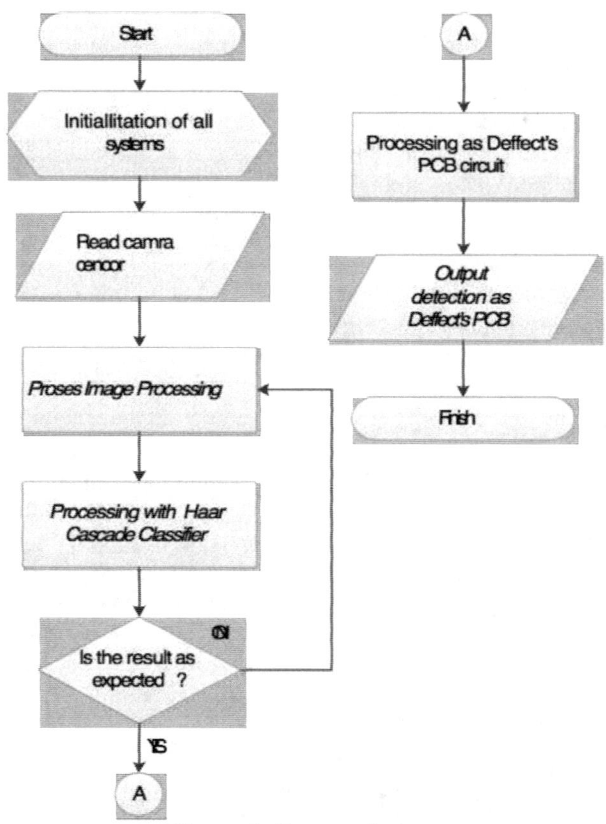

Figure 1. System design.

2.2. PCB's Deffect Detector Design

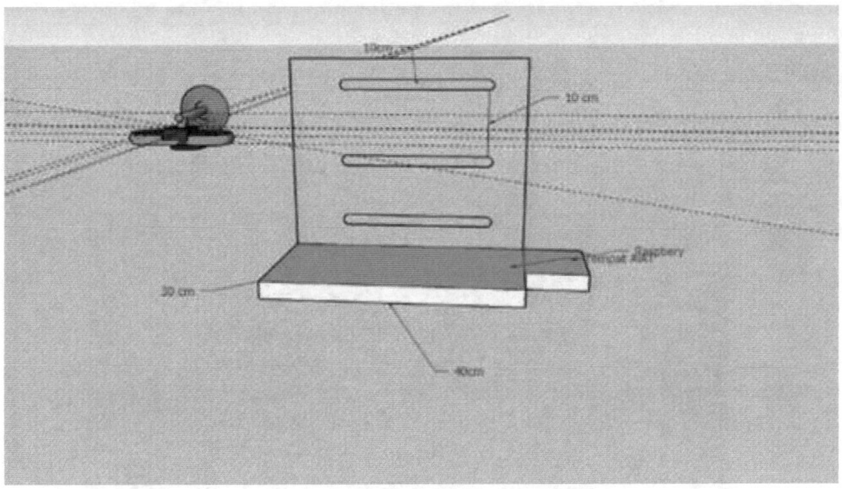

Figure 2. PCB's deffect detector design.

Figure 2 is the hardware design will be used on this research , it has 35 cm tall and there ae 3 lines that have 10 cm distance each, these lines use to put the camera censor. And there is a table with 30 cm width and 40 cm long, this table use to put the PCB that will be detected. And there is small surface beside the big table it used toput Raspberry Pi

2.3. Interfacing Software Design

At this stage what the author does is to use the Thonny software which is used to create views and input coding. Thonny software is used for coding assignments and running programs, as well as for system interfacing. In addition, the Thonny software imports several libraries that are used.

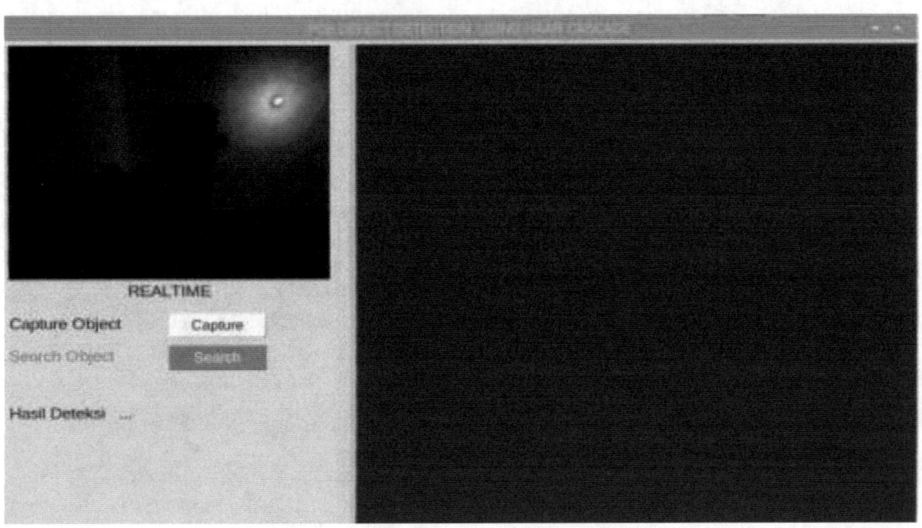

Figure 3. Interfacing Design

On this interfacing Design There are two windows, small windows and Big windows. The small window used to show the real time image from the PCB. The bigger window is shown PCB detection result as a picture . and capture button use to take the image of realtime PCB. Search button use to find the dataset that can be load to test the software work properly or not. And then hasil deteksi or detection result label use to show how many Detected defect circuit from the PCB.

3. Result and Discussion

3.1. Camera Testing Result

Camera is the main censor in this research especially for taking PCB's real time image. Author already testing the camera that already installed and programmed in to the Raspberry Pi.

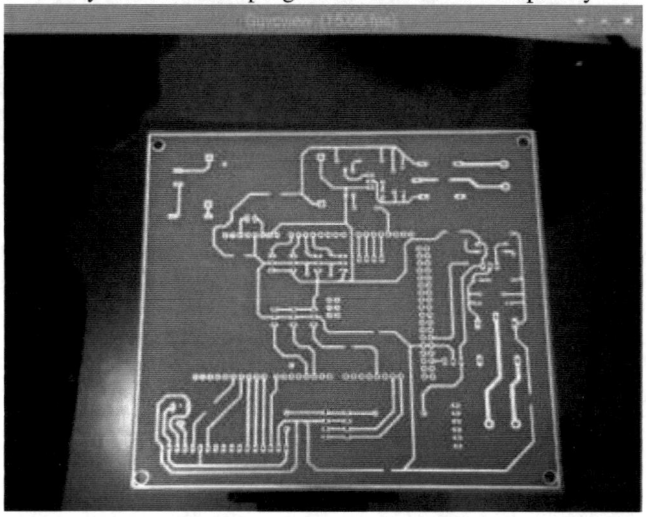

Figure 4. Camera testing.

3.2. Haar- Like Feature

This test was done by selecting the haaar-like feature in to square, in ths case the selecting feature are PCB's broken circuits. The first step in this test was changing the RGB image from PCB into Grayscale image so the system can deciding broken circuit feature.

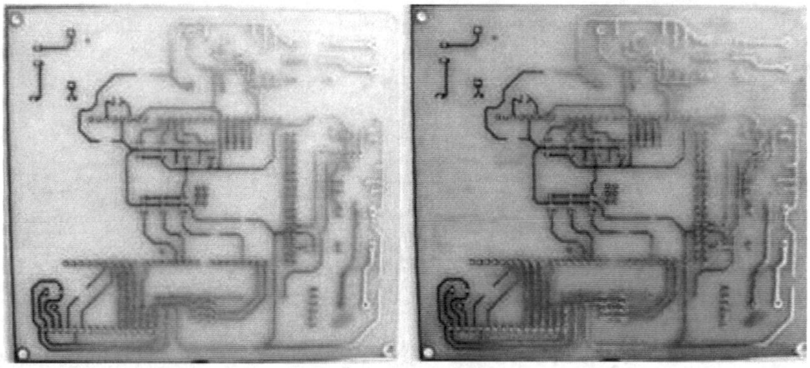

Figure 5. Convertion from RGB image to grayscale

IOP Conf. Series: Materials Science and Engineering 1175 (2021) 012011 doi:10.1088/1757-899X/1175/1/012011

After converting image to Grayscale image, the next step is to choose Haar-Like feature . Haar-Like Feature processing image in to squares , wich in one square there are some pixels. Than every square will be count and produce result a threshold, treshol is the difference that show dark and light part. Those value that will be processed on image processing.

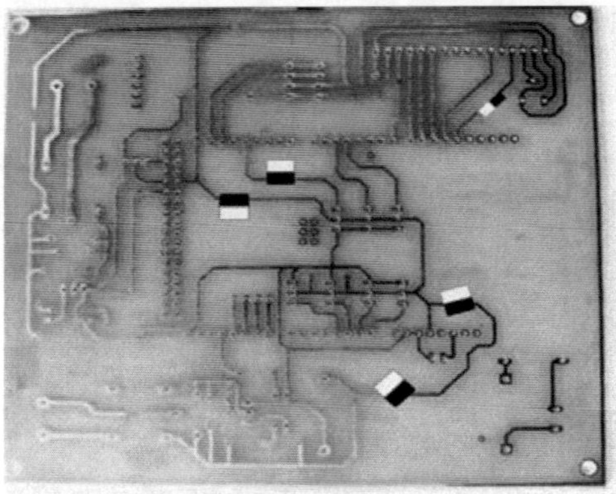

Figure 6. Haar-like feature

3.3. PCB's A Deffect Detection

In this test, the test will be carried out with a distance of 26 cm, 16 cm, and 6 cm, the distance which is the distance between the camera and the table where the PCB is placed, with each distance set in the camera angle of 60°, 90° and 105°. And before that author want to test if the system can detected a 0.5 mm broken circuit witdht or not, the result will be shown on table 1.

Table 1. Broken circuit with 0.5 and 1.5 mm width

PENGUJIAN PCB A PADA JARAK 26CM			
no	Lebar Jalur	Pengecekan konvensional	Pengecekkan menggunakanalat deteksi PCB dengan metode *Haar Cascade Classifier*
1.	1,5 mm	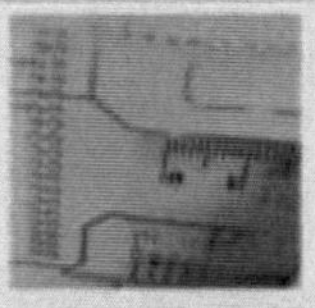	
2.	0,5 mm	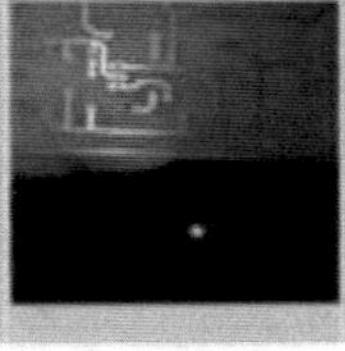	

From Table 1 one we can conclude that this tools can work as expected, after this PCB A will be tested in camera distance 26, 16, and 6 Cm with camera declivity angle 160°, 90° and 105°. The result show in table 2.

Table 2. PCB A Detection Result

	105°	90°	60°	Average
26 cm	97,5 %	72,5 %	52,5 %	74,16 %
16 cm	65 %	25 %	25 %	38,33 %
6 cm	Not deected	Not detect-ed	Not detect-ed	Not detect-ed
				52,49 %

3.4. PCB's B Deffect Detection

In this test, the test will be carried out with a distance of 26 cm, 16 cm and 6 cm, with each distance set in the angle of 60°, 90° and 105°. In PCB B this time the system can be considered running 100% when there is no PCB detection with broken paths.

Table 3. PCB B Detection Result

	105°	90°	60°	Average
26 cm	90 %	90 %	80 %	86,67 %
16 cm	97,5 %	92,5 %	95 %	95 %
6 cm	Not deected	Not detect-ed	Not detect-ed	Not de-tected
				90,83 %

4. Conclusion

The success rate of the system is influenced by the distance of the camera and also the position of the camera angle when taking images. It can be seen from each percentage that the average detection success rate is 97.5% for a distance of 26 cm, 38.83% for a distance of 16 cm, and cannot be detected at a distance of 6 cm. this is because the closer the camera position to the PCB, the narrower the field that can be taken and detected.

5. References

[1] Sulistyo dkk 2012 *Rancang Bangun Prototipe Aplikasi Pengenalan Wajah untuk Sistem Bsensi Alternatif dengan Metode Integral Feature dan Eigenface*

[2] Pradhana A and Prayudi, Y 2008 *Penggunaan Templete Matching untuk Identifikasi Kecacatan pada PCB*

[3] Sidharta H A 2017 *Introduction to Open CV* (Ilmu Komunikasi: Binus Malang, Oktober)
[4] Ahmad H 2010 *Haar Cascade Classifier.* https://www.researchgate.net/figure/HaarCascadeClassifier_fig1_330553730
[5] Jati 2015 Deteksi objek menggunakan *Haar Cascade Classifier* implemented. https://jati.stta.ac.id/2015/09/deteksi-obyek-menggunakan-haar-cascade.html
[6] lcdwiki.com 2019 3.5 inch Raspi display http://www.lcdwiki.com/3.5inch_RPi_Display
[7] Fadilla N 2017 Alat keselamatan pengendara dari kondisi lelah dilengkapi dengan *multi level safety* berbasis pengolahan citra dengan metode *Haar Cascade Classifier* (Tugas Akhir PPNS)
[8] Raspberrypi.org 2018 raspberrypi. [Available at: https://www.raspberrypi.org] Kementrian ESDM: Data Tumpahan Minyak 2017. [Online] Available: http://statistik.migas.esdm.go.id/index.php?r=dataTumpahanMinyak/index. [Accessed: 26-Jan-2020]
[9] Jami'in M A, Hu J, Marhaban M H, Sutrisno I and Mariun N B 2016 Quasi ARX neural network based adaptive predictive control for nonlinear systems. *IEEJ Trans Elec Electron Eng* **11** pp 83-90
[10] Jami'in M A, Sutrisno I and Hu J 2012 Lyapunov learning algorithm for Quasi-ARX neural network to identification of nonlinear dynamical system. *IEEE International Conference on Systems, Man, and Cybernetics (SMC)* (Seoul, 2012) pp 3147-3152 doi: 10.1109/ICSMC.2012.6378275
[11] Jami'in M A, Sutrisno I and Hu J 2013 Deep searching for parameter estimation of the linear time invariant (LTI) system by using Quasi-ARX neural network. *The 2013 International Joint Conference on Neural Networks (IJCNN)* (Dallas, TX) pp 1-5 doi: 10.1109/IJCNN.2013.6707091
[12] Jami'in M A, Sutrisno I and Hu J 2015 Maximum power tracking control for a wind energy conversion system based on a quasi ARX neural network model. *IEEJ Trans Elec Electron Eng* **10** pp 368-375
[13] Jami'in M A, Sutrisno I, Hu J, Mariun N B and Marhaban M H 2014 An adaptive predictive control based on a quasi-ARX neural network model. *13th International Conference on Control Automation Robotics & Vision (ICARCV)* (Singapore, 2014) pp 253-258 doi: 10.1109/ICARCV.2014.7064314
[14] Sutrisno I, Jami'in M A and Hu J 2014 Modified fuzzy adaptive controller applied to nonlinear systems modeled under quasi-ARX neural network. *Artif Life Robotics* **19** pp 22–26 (2014). https://doi.org/10.1007/s10015-013-0137-6
[15] Sutrisno I, Jami'in M A and Hu J 2014 An improved elman neural network controller based on quasi ARX neural network for nonlinear systems. *IEEJ Trans Elec Electron Eng* **9** pp 494-501
[16] Sutrisno I, Che C and Hu J 2014 An improved adaptive switching control based on quasi-ARX neural network for nonlinear systems. *Artif Life Robotics* **19** pp 347–353 (2014)
[17] Sutrisno I, Jami'in M A and Hu J 2012 Neural predictive controller of nonlinear systems based on quasi-ARX neural network. *18th International Conference on Automation and Computing (ICAC)* (Loughborough, 2012) pp 1-6
[18] Rinanto N and Khumaidi *Aplikasi Android- Rasspberry Pi pada Kapal Tanpa Awak untuk Pencarian Korban Kecelakaan Laut*
[19] Widodo R B et al Grasping and attached mode in human computer interaction in the study of mouse substitution. *Journal of Physics: Conference Series* **1196** 1 012034
[20] Sutrisno I, Jami'in M A, Hu J, Marhaban M H 2015 Self-organizing quasi-linear ARX RBFN modeling for identification and control of nonlinear systems. *54th annual conference of the society of instrument and control engineers of Japan (SICE).* IEEE
[21] Jami'in M A, Sutrisno I and Hu J 2014 Nonlinear adaptive control for wind energy conversion systems based on quasi-ARX neural networks model. *Procedings of the International Multi*

Conference of Engineers and Computer Scientists March 12-14, 2014 (Hong Kong, 2014)
[22] Sutrisno I 2009 *Pemrograman Komputer Dengan Software Matlab.* (Surabaya: ITS Press)
[23] Rinanto N and Khumaidi *Aplikasi Android-Rasspberry Pi pada Kapal Tanpa Awak untuk Pencarian Korban Kecelakaan Laut*

IOP Conf. Series: Materials Science and Engineering 1175 (2021) 012012 doi:10.1088/1757-899X/1175/1/012012

Application of Machine Learning Method to Predict Reliability in Lubricating Oil System Components

Nurvita Arumsari[1], Feby Agung Pamuji[2], Bianda Devi Puspita[3]

[1,3] Marine Engineering Department, Shipbuilding Institute of Polytechnic Surabaya, Surabaya, Indonesia

[2] Electrical Engineering Department, Institut Teknologi Sepuluh Nopember, Surabaya, Indonesia

E-mail: arum@ppns.ac.id, feby@ee.its.ac.id

Abstract. In predicting the reliability and failure of components, classical methods are often used by determining the distribution between failure times. Sometimes, the determination of this distribution does not always match the data pattern that is owned because of the limited data records and the many types of distribution that must be chosen. In addition, how much influence the time series has on components cannot be clearly analyzed. Therefore, in this study a prediction will be carried out by combining classical methods with machine learning, namely Support Vector Regression (SVR) and Least Square Support Vector Regression (LSSVR). Both methods are considered capable of improving the prediction accuracy of a series of data. The results showed that the classical combined method with LSSVR had better accuracy than SVR.

1. Introduction

In carrying out component maintenance or maintenance on the lubricating oil system or in the main engine, it is necessary to predict the appropriate time for maintenance or component replacement beforehand. The time in question is the failure time between components. Reliability is an opportunity for the component or system to fulfill a predetermined task without failing for a certain period of time if it is operated properly in a certain environment[1].

The reliability value pattern based on the failure time between components will be adjusted to the distribution of the data which will then be estimated using machine learning (ML). According to research conducted by [2], the ML method is considered to be more effective and accurate in estimating a continuous data. In machine learning data is needed as training before issuing output then the training results will be tested with the same or contradictory data. There are two main categories in machine learning, namely classification and prediction. There are many forecasting techniques that are applied in various studies. For example, to predict the life span of a component, the classical time series method is often used. However, the method is largely limited by the classification of failure types so that it is difficult to validate all the assumptions that exist [3]

Machine learning is an application of computer science, specifically Artificial Intelligence (AI), which provides automatic system performance without being explicitly programmed. The learning process begins with data observation or searching for data patterns and making better decisions. Machine learning algorithms have been around for a long time, with the ability to automatically apply complex mathematical calculations to big data and faster. Broadly speaking, machine learning techniques or ways

Content from this work may be used under the terms of the Creative Commons Attribution 3.0 licence. Any further distribution of this work must maintain attribution to the author(s) and the title of the work, journal citation and DOI.
Published under licence by IOP Publishing Ltd

of working are divided into 4 types, namely supervised learning, unsupervised learning, semi supervised learning, and reinforcement learning.

The application of machine learning in this study focuses on supervised learning by comparing two methods between Support Vector Regression (SVR) and Least Square Support Vector Regression (LS-SVR). The purpose of using both methods is to predict the reliability value of lubricating oil system components. The two methods are then hybridized by obtaining the results of the distribution of the failure data parameters generated through the software. The research object was selected from the lubricating oil system which has the most frequent interval between failure times in the last 5 years, namely Lubrication Oil (LO) Cooler and Strainer.

Both ML methods are the development of the Support Vector Machine (SVM) for numerical data types in the regression case. In the SVR algorithm, the response variable has a range in the form of real number values with the limitation function in the form of inequality [4]. Meanwhile, LS-SVR is a modification of SVR for numerical data with a boundary function in the form of an equation.

1.1 Lubricating Oil System

The following is a figure of the Lubricating Oil System and how it works

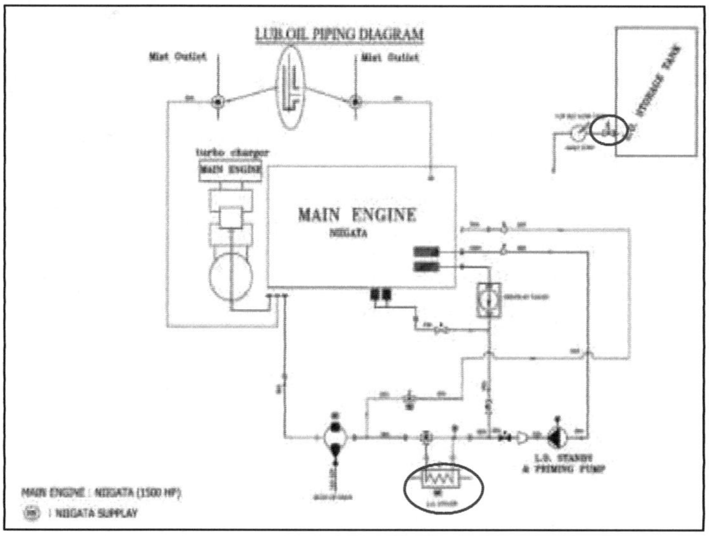

Figure 1. Lubricating Oil System With Strainer and LO Cooler Component

The lubrication system used is a wet sump with the type of main with the engine type is niigata. The principle of work start from the process of supplying lubricating oil from the storage tank to the sump tank with the help of a gear type hand pump. Before the lubricating oil flows into the main diesel engine, the lubricating oil will pass through the strainer in the sump tank for the filtering process, and then it will be pumped with a lube oil pump to the lube oil cooler. The temperature of the lubricating oil coming out of the cooler is automatically controlled at a constant level determined to obtain the desired viscosity of the main diesel engine. Then the lubricating oil will be divided into two lines, some of the lubricating oil will be flowed to the top of the main diesel to lubricate the components on the head such as the rocker arm, camshaft, etc. At the same time, some of the other lubricating oil will flow to the lube oil filter for the filtering process for the separation of lubricating oil. With water and unwanted substances. Furthermore, the lubricating oil will go to the main engine bearing and also flow back to the sump tank

1.2 Support Vector Regression (SVR)

The SVR method aims to minimize risk so as to be able to overcome cases that are overfitting. The general function of the SVR is as follows [5].

$$\mathrm{f}(X) = w^T \varphi(X) + b \tag{1}$$

where w is the weight vector, $\varphi(X)$ is a function that maps x to a dimension, and b represents the error value. Meanwhile, to minimize the risk the following functions are used.

$$R_{emp}(f) = \frac{1}{N}\sum_{i=1}^{N} \emptyset_\varepsilon\,(Y_i, f(X_i) + b) \tag{2}$$

Turns into an equation

$$R_{emp}(f) = \frac{1}{N}\sum_{i=1}^{N} \emptyset_\varepsilon\,(Y_i, W^T\varphi(X_i) + b) \tag{3}$$

Where $\emptyset_\varepsilon\big(Y, f(X)\big)$ is the ε-insensitive loss function which is defined as a function

$$\begin{cases} |f(X) - Y| - \varepsilon, & jika|f(X) - Y| \geq \varepsilon \\ 0, & untuk\ yang\ lain \end{cases} \tag{4}$$

The focus of SVR is to find the optimum hyperplane and minimize errors between training data and ε-insensitive loss as shown in the following equation.

$$\min_{W,b,\xi^*,\xi} R_\varepsilon(W, \xi^*, \xi) = \frac{1}{2}W^T W + C\sum_{i=1}^{N}(\xi_i^* + \xi_i) \tag{5}$$

With limitations

$$\begin{gathered} Y_i - W^T\varphi(X_i) - b \leq \varepsilon + \xi_i^*, \quad i = 1,2,..,N \\ -Y_i + W^T\varphi(X_i) - b \leq \varepsilon + \xi_i, \quad i = 1,2,..,N \\ \xi_i^* \geq 0 \\ \xi_i \geq 0 \end{gathered}$$

C is defined as a parameter with a maximum tolerance limit. Whereas W is a parameter vector with equations

$$W = \sum_{i=1}^{N}(\beta_i^* - \beta_i)\,\varphi(X_i) \tag{6}$$

By using the quadratic and lagrange program solving function, the following SVR final model is obtained.

$$f(X) = \sum_{i=1}^{N}(\beta_i^* - \beta_i)K(X_i, X) + b \tag{7}$$

With limitations

$$\sum_{i=1}^{N}(\beta_i^* - \beta_i) = 0; 0 \leq \beta_i \leq C; 0 \leq \beta_i \leq C$$

And $K(X_i, X)$ is a kernel function.

IOP Conf. Series: Materials Science and Engineering 1175 (2021) 012012 doi:10.1088/1757-899X/1175/1/012012

1.3 Least Square Support Vector Regression (LS-SVR)

The LS-SVR method was first developed by [6]. LS-SVR is a modification of the SVR method with the same goal of minimizing risk. This LS-SVR limitation function is in the form of an equation so it is predicted that it will produce a unique and more efficient solution. The objective function of the LS-SVR is shown the same as the SVR as follows [7].

$$f(X) = w^T \varphi(X) + b$$

Based on the principle of minimizing risk, the optimum function and limits of the LS-SVR are as follows.

$$min\, J(w,e) = \frac{1}{2} w^T w + \frac{1}{2} c \sum_{i=1}^{N} e_i^2 \quad (8)$$

$$s.t \quad y_k = w^T \varphi(x_i) + b + e_k, \quad k = 1,2,\dots,N \quad (9)$$

Where J is the loss function and C is the maximum tolerance limit. With the lagrange equation, the optimum function is transformed as follows.

$$L(w,b,e,\alpha) = J(w,e) - \sum_{k=1}^{N} \alpha_k \{w^T \varphi(x_k) + b + e_k - y_k\} \quad (10)$$

With α_k are Lagrange multipliers.
In LS-SVR, the optimization problem can be simplified by solving linear equations rather than the more complex SVR. Thus, the final LS-SVR model is given as follows.

$$y(x) = \sum_{k=1}^{N} \alpha_i K(X, X_k) + b \quad (11)$$

Where $K(X_i, X)$ is the kernel function and α_i is in vector form.

1.4 Time To Failure Distribution

To measure the Reliability R(t) of each components, first determine the type of distribution and its parameters. The following is the formula of the 2 parameter exponential and 2 parameter Weibull distribution [14].

The formula of R(t) 2 parameter exponential distribution

$$R(t) = exp\left[-\frac{x-\gamma}{\theta}\right], x > \gamma, \theta > 0, -\infty < \gamma < \infty \quad (12)$$

The formula of R(t) 2 parameter weibull distribution

$$R(t) = exp\left[-\left(\frac{x}{\alpha}\right)^{\beta}\right], x > \gamma, \theta > 0, -\infty < \gamma < \infty \quad (13)$$

With
θ is scale parameter
γ is threshold parameter
α is scale parameter
β is shape parameter

1.5 Goodness of Fit Test Distribution

Determination of the distribution of Time to failure (TTF) components is based on the output Goodness of Fit value. In this case, the calculation of the value of Goodness of Fit uses the Anderson - Darling statistic [8].

Let $Z = F(X)$, where $F(X)$ is the cumulative distribution function (CDF). Suppose that a sample $X_{1,....,}X_n$ gives values $Z_{(i)} = F(X_i),\ i = 1, \dots, n$.
Rearrange $Z_{(i)}$ in ascending order $Z_{(1)} < Z_{(2)} < \cdots < Z_{(n)}$.

Then, the Anderson-Darling Statistic (A^2) is calculated as follows.

$$A^2 = -n - \left(\frac{1}{n}\right)\sum_i\left[(2i-1)logZ_{(i)} + (2n+1-2i)log\left(1 - Z_{(i)}\right)\right] \quad (14)$$

Besides Anderson-Darling statistics, goodness of fit test distribution is also determined by p-value exact.

1.6 Evaluation of Predicted Results

In evaluating the accuracy of the prediction results of reliability, two types of errors are used, namely MAE (Mean Absolute Error) and RMSE (Root Mean Square Error) [9] which are stated as follows.

$$MAE = \frac{\sum_{n=1}^{N}|\widehat{r_n} - r_n|}{N} \quad (15)$$

$$RMSE = \sqrt{\frac{\sum_{n=1}^{N}(\widehat{r_n} - r_n)}{N}} \quad (16)$$

Where $\widehat{r_n}$ is the predicted value, r_n is the observed value and N is the total predicted data.

1.7 The Discussion about Reliability Prediction Research

Basically, many researches on reliability forecasting have been done. One of the studies on reliability was carried out by [10] who predicted the reliability concrete of components using conventional methods through determining the Weibull parameter distribution based on experimental data. Further research that combines reliability with machine learning methods, among others, was conducted by [11] and [12].

The two researchers used ridge regression and support vector regression methods in predicting the component reliability values. In ridge regression, the predicted data pattern has a linear pattern, while in SVR it can be used to predict non-linear data patterns because it involves kernel functions. In this study, the degenerate data in the SVR method was failure time data so that the results obtained were only nonparametric values without knowing the appropriate distribution of failure time data. This study tries to combine the results obtained from conventional methods and machine learning, both SVR and LSSVR. Research that has been conducted by [2] and [13] has proven that the SVR method is better at estimating non-linear patterns with the RBF kernel model. Meanwhile, according to [7] the LSSVR method is considered to have a more accurate prediction result.

2. Research Methodology

The data used in this study are data on the failure time of the LO Cooler and strainer components in the lubricating oil system with a numerical scale. Data on component failure times are in the form of data series for the last 5 years from 2013 to 2018. As previously explained, the proposed method in this study is to combine conventional methods with machine learning.

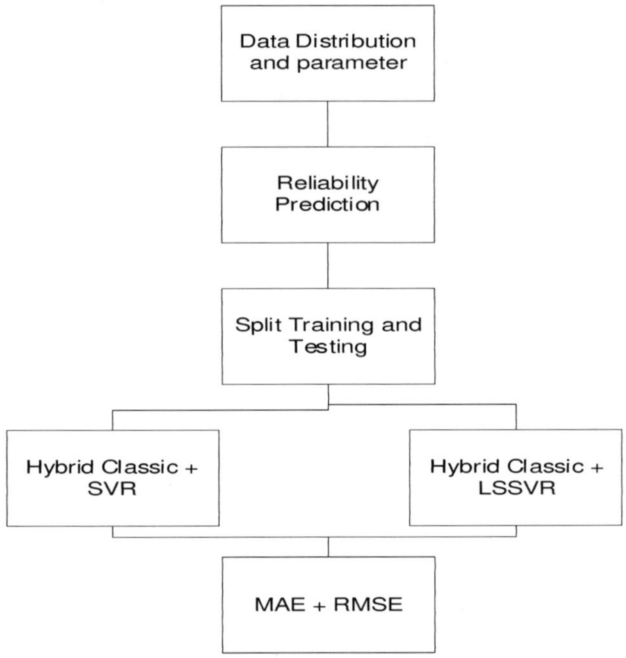

Figure 2. Research Flowchart

3. Result and Discussion

In applying the machine learning method to the LO Cooler and Strainer components, the parameter distribution will first be determined. The parameter distribution is determined based on the failure time data record between components. The results are stated in Table 1 below.

Table 1. Data Distribution and Parameter

Component	Distribution	Shape	Scale	Threshold
LO Cooler	2 Parameter Eksponential	-	2057, 08	3558,87
Strainer	2 Parameter Weibull	10,92	3503, 4	-

From the distribution parameter results obtained in Table 1, then the reliability prediction for each component with formula 12 and 13 are carried out at certain time intervals.

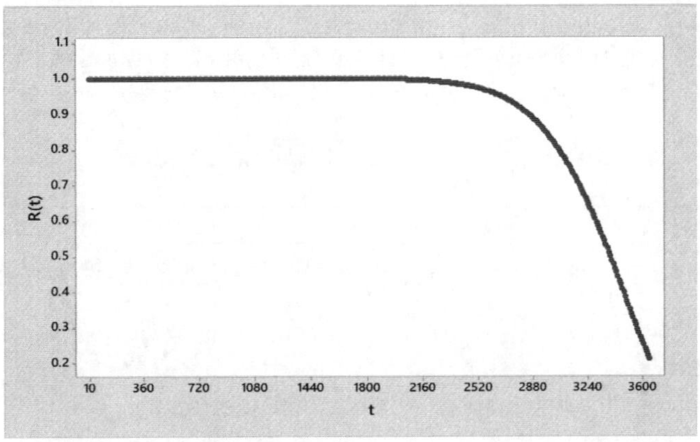

Figure 3. Reliability Prediction of Strainer Component

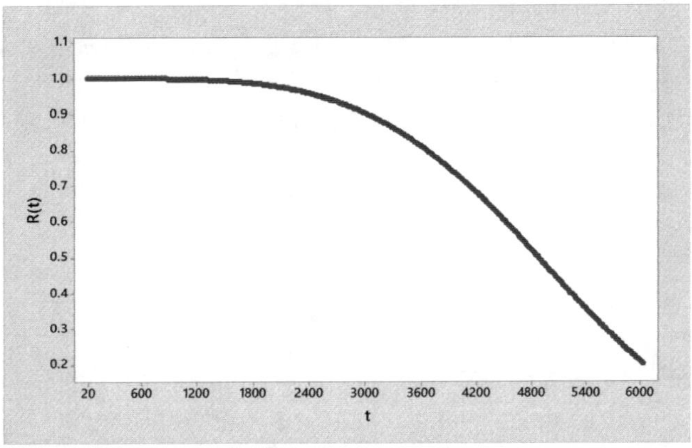

Figure 4. Reliability Prediction of LO Cooler Component

Based on Figures 3 and 4, it can be seen that the reliability pattern of LO Cooler tends to sharpen after t = 3000 hours, while in the strainer after t = 2880 hours. This shows that the strainer component fails more frequently than the LO Cooler. In the Strainer, after a time interval of 3650 hours the component has a reliability value of less than 0.2 and ideally component replacement should be carried out. Meanwhile, if you want to treat the strainers, it should be done at intervals of every 3290 hours or about 4.5 months. This is because after t the reliability value of the strainer is less than 0.6.

Meanwhile, the LO Cooler has a less frequent failure than the strainer. In LO Cooler, component replacement should be done after t = 6040 hours because it has a reliability below 0.2. Meanwhile, if you want to do maintenance on the LO cooler, it should be done at intervals of 4500 hours or about 6 - 6.5 months. This is because, after an interval of 4500 hours the LO cooler has a reliability of less than 0.6.

From the prediction results of reliability with conventional methods, it can be seen that the pattern formed is non-linear because it follows an exponential and weibull pattern. Thus, the next recommended method is a method that is able to overcome the nonlinearity problem so that it can produce more accurate predictions. In this study, the two recommended methods for hybridization with conventional methods are SVR and LSSVR.

After obtaining the predictive value of classical reliability, namely by adjusting the distribution patterns and parameters obtained, then it will be hybridized with the SVR and LSSVR machine learning methods. Before prediction with the two methods, first the data reliability will be displaced into training and testing data. In this study, the percentage of training data selection was 90% and testing was 10%.

The kernel function was selected based on the results of previous research, namely the RBF (Radian Basis Function) Kernel. The selection of the RBF kernel function is considered the most appropriate in estimating non-linear patterns.The prediction results of reliability can be seen in Table 2 below.

Table 2. Results of the Reliability Prediction

t	Actual		Hybrid SVR		Hybrid LSSVR	
	Cooler	Strainer	Cooler	Strainer	Cooler	Strainer
1	0.3401	0.6044	0.3371	0.6103	0.3400	0.6045
2	0.3348	0.5943	0.3322	0.6107	0.3347	0.5942
3	0.3295	0.5839	0.3274	0.5932	0.3294	0.584
4	0.3243	0.5735	0.3226	0.585	0.3242	0.574
5	0.319	0.563	0.318	0.5769	0.3191	0.5631
6	0.3139	0.5523	0.3134	0.569	0.3138	0.552
7	0.3087	0.5416	0.3089	0.5613	0.3088	0.5415
8	0.3036	0.5307	0.3045	0.5539	0.3035	0.5306
9	0.2985	0.5197	0.3002	0.5468	0.2984	0.5198
10	0.2934	0.5087	0.296	0.5401	0.2935	0.5086

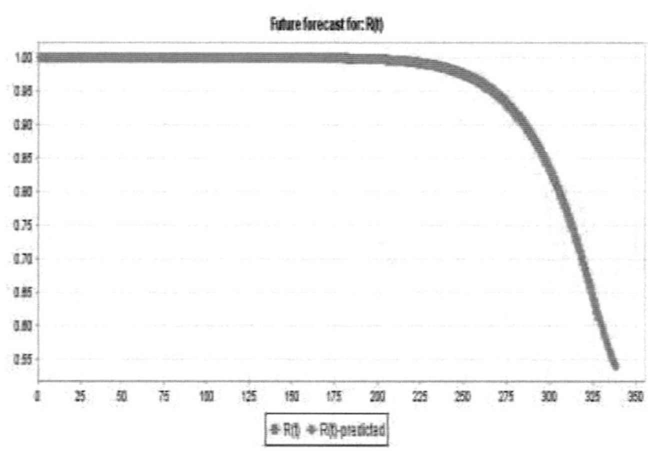

Figure 5. Reliability Prediction of Strainer Component with Hybrid SVR

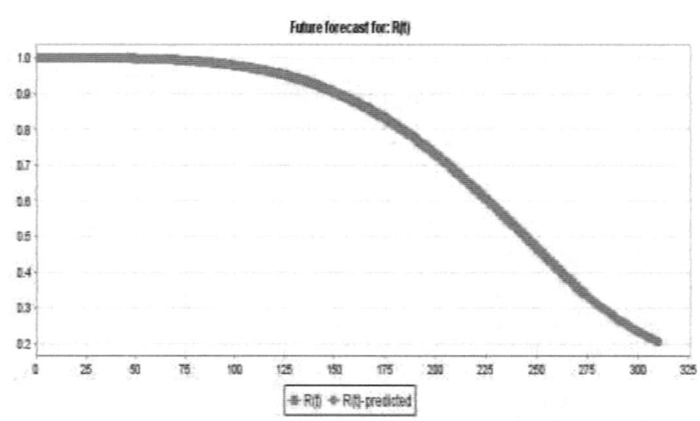

Figure 6. Reliability Prediction of LO Cooler Component with Hybrid SVR

Based on the results of the analysis of the reliability prediction with hybrid SVR for the Strainer and LO Cooler components in the lubricating oil system, it is known that the predictive value obtained by using training data is almost close to the true value. In this case, the reliability value is actually the result of conventional prediction by generating distribution. Based on Figures 5 and 6, it can be seen that the prediction results follow the reliability distribution graph pattern. In the strainer component, the reliability with the observation index above 250 has a fairly sharp pattern of decline, while in the LO Cooler, a significant decrease in the predicted number occurs after the 175th observation index. In this case, the observation index does not represent the actual time. The time interval given can differ according to the needs of the interpretation of the researcher. A sharp decrease in the number of reliability indicates that the use of components is predicted to exceed the lifespan. The prediction results of the two components of the lubricating oil system are then compared with the LSSVR hybrid method as shown in Figures 7 and 8 below.

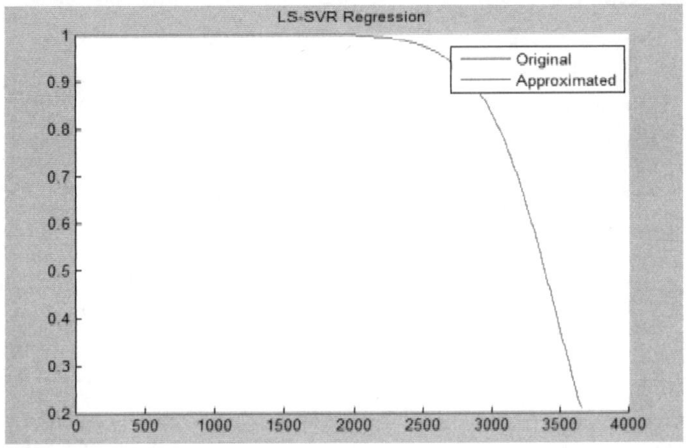

Figure 7. Reliability Prediction of Strainer Component with Hybrid LSSVR

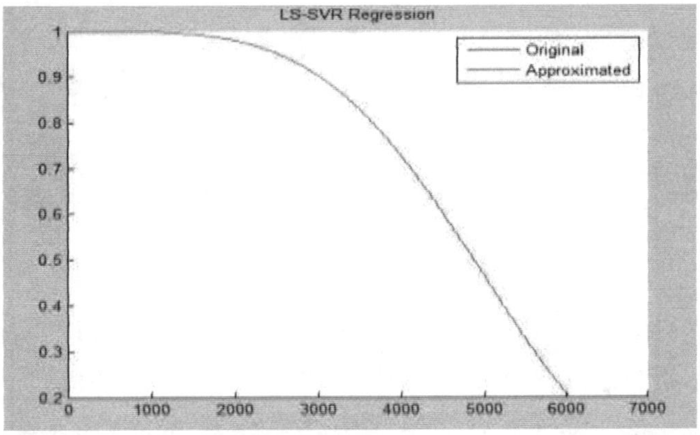

Figure 8. Reliability Prediction of LO Cooler Component with Hybrid LSSVR

The results of descriptive analysis through graphic observations for the two components of the lubricating oil system show that the predictive value of the reliability obtained closely coincides with the actual predicted value. The axes show the time intervals in hours. Descriptively, when compared with the predicted numbers using the hybrid SVR method, it can be seen that the hybrid LSSVR method tends to be better because it is very similar to the original pattern of the actual prediction results.

IOP Conf. Series: Materials Science and Engineering 1175 (2021) 012012 doi:10.1088/1757-899X/1175/1/012012

Finally, the following of contrast error between actual and prediction result with Hybrid SVR and LSSVR can be see in Figure 9 and 10

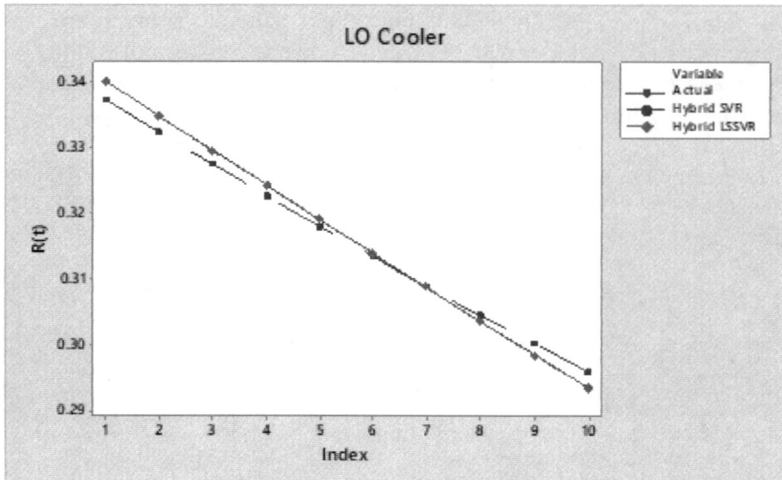

Figure 9. The Contrast Error of Reliability Prediction LO Cooler

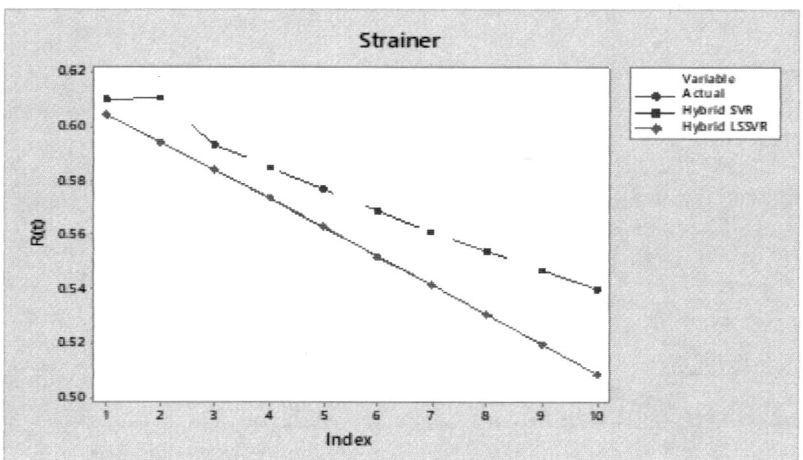

Figure 10. The Contrast Error of Reliability Prediction Strainer

To ensure a descriptive description of the results of the analysis of the two recommended methods, model validation was carried out by calculating the selected errors, namely MAE and RMSE, as shown in Table 3. The calculation of the error value comes from the prediction results with training data which is then compared with the testing data. Overall validation results show that the hybrid LSSVR method is better than the hybrid SVR. This is shown in the error acquisition of the two components using both MAE and RMSE which is smaller than the hybrid SVR. Meanwhile, when compared between components using both methods, it can be seen that the LO cooler has a smaller error value compared to the Strainer.

Table 3. Error Validation With Hybrid SVR dan LSSVR

Component	Hybrid SVR		Hybrid LSSVR	
	LO Cooler	Strainer	LO Cooler	Strainer
MAE	0.00948	0.1005	0.000019	0.000011
RMSE	0.0022	0.1254	0.00000197	0.0000019

4. Conclusion

In this study, the application of predictive method selection is a method commonly used for nonlinear cases. In this case, the SVR and LSSVR methods were chosen because in addition to being compatible with the nonlinear case, the two methods were considered to be able to improve the accuracy of the component prediction values. The choice of kernel function in this study is based on the results of previous studies which correspond to the nonlinear case, namely the RBF kernel. The contribution given to this paper is in the form of combining classic reliability prediction methods with machine learning. The results obtained, the combined method can improve the prediction results of component reliability, especially in LSSVR. For further research, we can apply the Hybrid LSSVR and SVR method to conduct predictive maintenance in LO Cooler component.

Acknowledgements

The author would like to thank those who cannot be named one by one (anonymous) who have provided various suggestions and references for a significant improvement in paper quality.

References

[1] H. Al Haiany 2016"Reliability Centered Maintenance Different Implementation Approaches," *Eng. Maint.*

[2] E. K. Adhitya, R. Satria, and H. Subagyo 2015 "Komparasi Metode Machine Learning dan Metode Non Machine Learning untuk Estimasi Usaha Perangkat Lunak," *J. Softw. Eng.*, vol. 1, no. 2, pp. 109–113.

[3] Y. H. Liang 2011 "Analyzing and forecasting the reliability for repairable systems using the time series decomposition method," *Int. J. Qual. Reliab. Manag.*

[4] A. J. Smola and B. Schölkopf 2004 "A Tutorial on Support Vector Regression. Statistics and Computing archive," *Stat. Comput.*

[5] G. F. Fan, L. L. Peng, W. C. Hong, and F. Sun 2016 "Electric load forecasting by the SVR model with differential empirical mode decomposition and auto regression," *Neurocomputing.*

[6] J. A. K. Suykens, T. Van Gestel, J. De Brabanter, B. De Moor, and J. Vandewalle 2002 "*Basic Methods of Least Squares Support Vector Machines*," in Least Squares Support Vector Machines.

[7] G. Y. -, Z. L. -, L. X. -, R. C. -, and M. J. 2013 "A New Time Series Prediction Method Based on LS-SVR and AGO," Int. J. Adv. *Comput. Technol.*, vol. 5, no. 6, pp. 542–551.

[8] L. N. Harris 1987 "*Goodness-of-fit techniques, Ralph B. D'Agostino and Michael A. Stephens (eds).* Marcel Dekker Inc. 1986. No. of pages: 560. Price: $79.75 (U.S.A. and Canada); $95.50 (elsewhere)," Qual. Reliab. Eng. Int.

[9] W. Wang and Y. Lu 2018 "Analysis of the Mean Absolute Error (MAE) and the Root Mean Square Error (RMSE) in Assessing Rounding Model," in IOP Conference Series: *Materials Science and Engineering.*

[10] Q. S. Li, J. Q. Fang, D. K. Liu, and J. Tang 2003 "Failure probability prediction of concrete components," *Cem. Concr. Res.*, vol. 33, no. 10, pp. 1631–1636.

[11] M. D. C. Moura, E. Zio, I. D. Lins, and E. Droguett 2011 "Failure and reliability prediction by support vector machines regression of time series data," *Reliab. Eng. Syst. Saf.*, vol. 96, no. 11, pp. 1527–1534.

[12] S. Mangalathu and J. S. Jeon 2018 "Classification of failure mode and prediction of shear strength for reinforced concrete beam-column joints using machine learning techniques," *Eng. Struct.*, vol. 160, no. September 2017, pp. 85–94.

[13] N. Arumsari and F. A. Pamuji 2017 "Peramalan Irradiance Cahaya Matahari pada Sel Surya untuk Memenuhi Kebutuhan Energi Listrik dengan Metode Support Vector Regression (SVR)," *J. Nas. Tek. Elektro*, vol. 6, no. 1, p. 18.

[14] https://support.minitab.com/en-us/minitab/18/help-and-how-to/quality-and-process-improvement/quality-tools/how-to/individual-distribution-identification/methods-and-formulas/distributions/. **35** L74

Monitoring system on towercopter using ELM (Extreme Learning Machine) method

A D Santoso[1], L Nurkhamidah[2], Tibyani[3], V Y P Ardhana[4], I Sutrisno[5], Z M A Putra[6], M S Kahar[7]

[1]Politeknik Pelayaran Surabaya, Indonesia
[2,5,6]Politeknik Perkapalan Negeri Surabaya, Indonesia
[3]Universitas Brawijaya Malang, Indonesia
[4]Universitas Qamarul Huda Badaruddin, Indonesia
[7]Universitas Muhammadiyah Sorong, Indonesia

E-mail: agusbp2ipsby@gmail.com

Abstract. Towercopter has experienced rapid development in recent years. Towercopter is a tool with one brushless motor and one propeller that is installed between two support poles. Towercopter requires balance to be able to move during takeoff or houvering. In this final project, a study will be developed using three brushless motors. Towercopter is a singlecopter prototype that specifically moves vertically. The load disturbance test on the system causes the towercopter to become unstable while working towards the setpoint. It needs to be controlled so that it is stable. So to solve these problems a prototype is made. Towercopter in this final project is a prototype designed using three brushless motors and 3 propellers which are mounted on two support poles, namely stainless axles. Using two sharp irradi sensors to detect height and slope. The purpose of this final project is to produce an altitude control on a towercopter using the ELM (Extreme Learning Machine) method.

1. Introduction

Looking at the geographical position of Indonesia which is located at the end of the movement of three world plates, namely Eurasia, Indo-Australia, and Pacific, making Indonesia a cataclysm-prone country [1]. The impact caused by the cataclysm is detrimental to the surrounding community. Besides the loss of material, the evacuation process, victim's search, and disaster management process frequently also take a long time due to the condition of the disaster is difficult to access [1].

In the process, evacuation and monitoring of cataclysm locations can be carried out through land and Air. Whereas through the air, usually use the assistance of helicopters to comb the location of the disaster (I Gede Andika, 2018).A Helicopter one of a kind of Unmanned Aerial Vehicle (UAV) which is an unmanned aircraft that does not require a human operator, can fly independently, or be driven remotely and can carry cargo [2].

However, in fact, in the field some problems are natural disturbance in the area is bigger like wind and also because the burden itself when the helicopter dropped the goods or raise the goods using a rope with the help of the pilot so that the stability in the helicopter bad. To improve stability, a height controller is required to increase the stability of the helicopter when carrying the load.

Content from this work may be used under the terms of the Creative Commons Attribution 3.0 licence. Any further distribution of this work must maintain attribution to the author(s) and the title of the work, journal citation and DOI.
Published under licence by IOP Publishing Ltd

IOP Conf. Series: Materials Science and Engineering 1175 (2021) 012013 doi:10.1088/1757-899X/1175/1/012013

The previous research was conducted by Chintia Putri Unarya under the title "Vertical motion arrangement System Planning in Towercopter using Labview-based control PID" [3]. In this study the method still has drawbacks in terms of the time it takes to get the PID parameter and still use the ultrasonic sensor HC-SRO4 which causes the sensor to still be less precise and tends to be unstable so that the resulting response is less good.

The previous research was conducted by Chintia Putri Unarya under the title "Vertical motion arrangement System Planning in Towercopter using Labview-based control PID" [3]. In this study the method still has drawbacks in terms of the time it takes to get the PID parameter and still use the ultrasonic sensor HC-SRO4 which causes the sensor to still be less precise and tends to be unstable so that the resulting response is less good.

Rifqi Nabila Zufar also conducts a study under the title "Design-build of altitude position control system on Proportional-based Integral Towecopter using Artificial Neural Network Tuning" In this study also still uses the ultrasonic sensors HC-SR04 that the sensory readings are still less precise and can be developed using the PID-ANN method to reduce existing oscillation. In both studies, it still uses 1 input and 1 output.

Therefore, the purpose of this research is to create a design-build towercopter with 4 inputs and 3 outputs. Equipped with 2 IR sharp sensors that can add sensor precision in the data readings and 2 load cell sensors to add interference test to the system of load, and 3 brushless motors as Actuators. This study using the ELM method is expected to be able to repair two previous methods. Therefore, a study titled "Design-Build of altitude position control system on Towercopter using ELM (Extreme Learning Machine) method".

2. Methodology

2.1.System Design

Flowchart System is a working step of the system, namely system initialization first by making sure all hardware is connected and can work well. After the initialization of the system is to determine the set point for the height of the towercopter. After that Towercopter will work towards SetPoint. Then the altitude on the towercopter will be read by the sharp IP sensor. Afterward plus a test of interference by adding the sensor load cell and slope detected by the IR sharp sensor. Some of these inputs are controlled by the ELM process to deliver the output to send commands to the brushless motor to move the propeller to the set point on the Towertreecopter. If the system has not reached the setpoint then the system returns to the ELM process. When the system has reached a set point then the data will be monitored by Android and to display the chart set point.

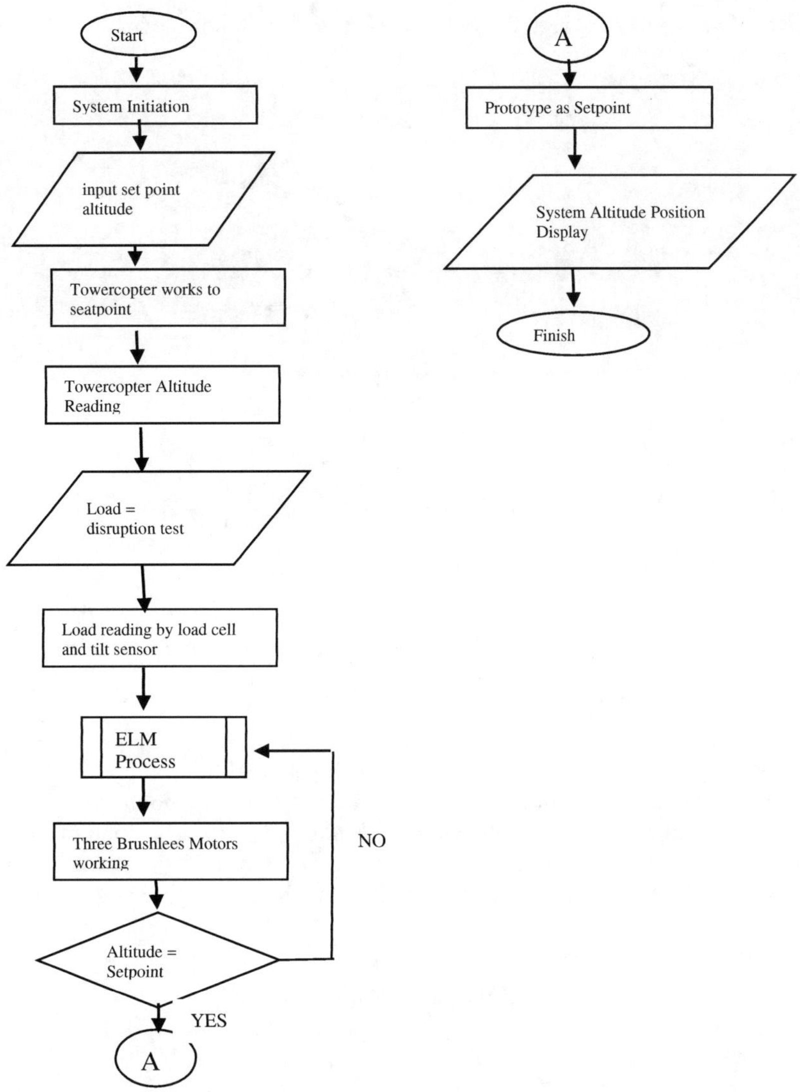

Figure 1. System Design.

IOP Conf. Series: Materials Science and Engineering 1175 (2021) 012013 doi:10.1088/1757-899X/1175/1/012013

2.2. Towercopter Design

The following is a hardware design design that will be used in this final project.

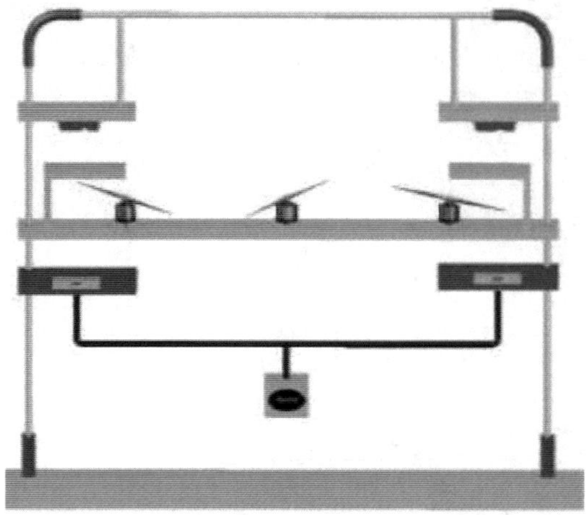

Figure 2. Towercopter Hardware Design.

The mechanical design of the tool in this final project has been illustrated in the hardware design which is equipped with three drivers and four sensors so that the system can run as intended.

2.3. ELM Model Design

From Figure 3 can be explained that this final task uses 4 inputs and 3 outputs. The ELM itself is an artificial neural network that uses one hidden layer. There are 4 inputs symbolized by x1, x2, X3, and X4 that display both the altitude and weight values by both sensors. The Output is a speed of 3 brushless motors symbolized by T1, T2, T3. The ELM structure above consists of Matrik H, Matrik Beta, and Tansig used for the activation function. The ELM method was applied to adjust the propeller height of the towercopter to balance.

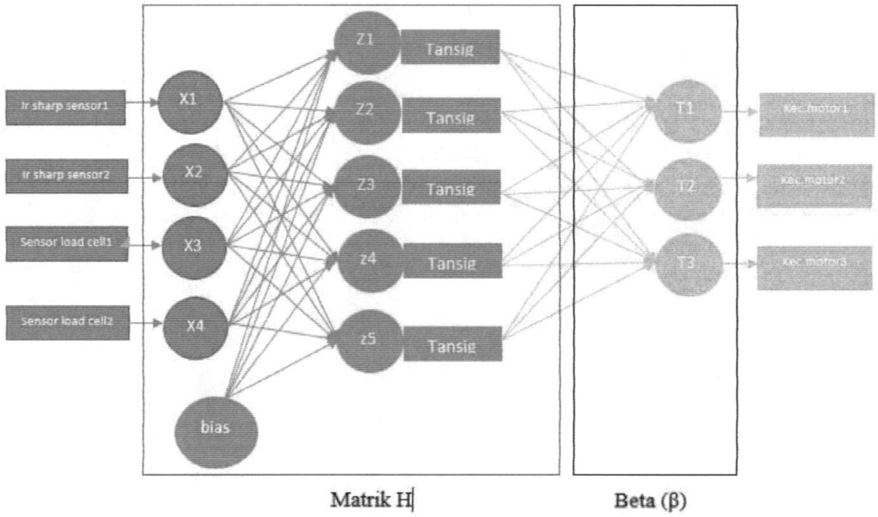

Figure 3. ELM Model Design

2.4. Android App Design

Based on figure 4 the fuction of this android is the control start and stop on towercopter. And to display the setpoint graph on the system. Setpoint control and also for monitoring sensor data namely height, load weight, and slope. Based on figure 15 the data is obtained from the tool, namely when inputting setpoint 50 wiil display the graph and data as in the android image below.

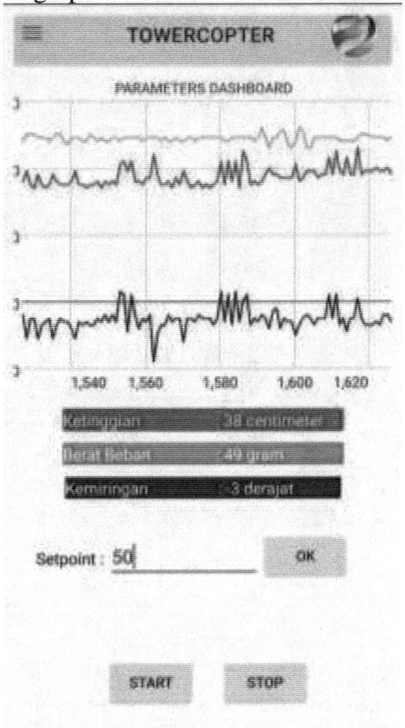

Figure 4. Android Application Loading Page.

Figure 5. Android Application Menu Page.

IOP Conf. Series: Materials Science and Engineering 1175 (2021) 012013 doi:10.1088/1757-899X/1175/1/012013

3. Result and Discussion

3.1. Towercopter Design Result

The following is the design for the placement of hardware on the mechanics shown in the image below.

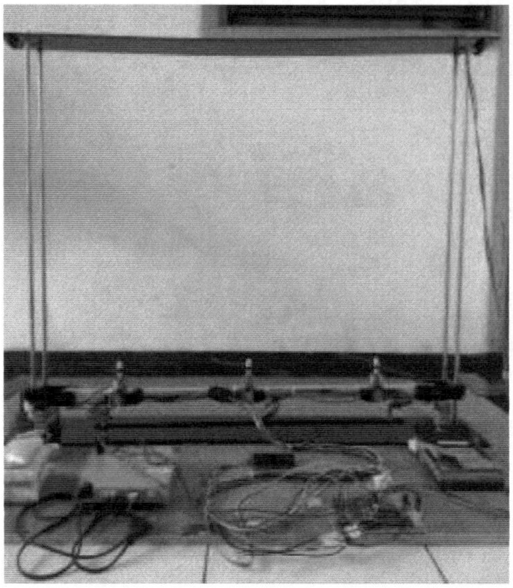

Figure 6. Towercopter design result.

The mechanical design in this final project uses two 8mm diameter stainless axles with 80cm length and 93cm 5mm acrylic cutting as the main frame. In making this mechanical hardware, it requires aluminum to put 3 brushless motors and 2 load cell sensors. Two linear bearings are attached to each stainless axle to drive the towercopter. On aluminum, a connecting road is installed to connect the load cell 1 sensor and the load cell 2 sensor on this mechanical hardware. In this tool, four sensors are installed, namely 2 ir sharp sensors and 2 load cell sensors. Ir sharp sensors are used to detect the height and slope of the towercopter. Which of the sensors is able to detect up to 80 cm while for this tool the maximum is up to 70 cm. The load cell sensor functions to detect the load to be tested for interference with the towercopter. In this mechanical design, there are actuators, namely 3 brushless motors that will be connected to the propeller.

3.2. Ir Sensor Sharp Testing

At this final task use 2 Ir sharp sensors that serve to detect altitude on the towercopter and tilt. At the test this time using a ruler workpiece that is. In this sensor test, the observed value is output distance that reads by the sensor with a distance range of 10 cm – 42 cm.

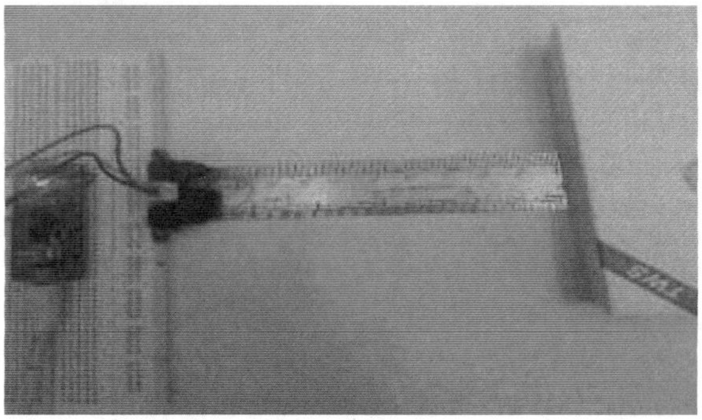

Figure 7. Ir sensor sharp Test.

Table 1. Ir sensor sharp testing data

Testing	Ruler(cm)	Ir sensor Sharp(cm)	Error (%)
1	10.00	10.07	0.7 %
2	17.00	17.35	2.05 %
3	20.00	20.48	2.4 %
4	21.00	21.56	2.6 %
5	23.00	23.64	2.7 %
6	26.00	26.42	1.61%
7	30.00	30.47	1.56 %
8	33.00	33.59	1.78 %
9	39.00	39.02	0.05 %
10	42.00	42.09	0.21 %
	Error average		= 1.5 %

The reading error of the sharp ir sensor is quite small, namely 1.56%. This test shows that the sensor has a good distance reading capability with very low error. Of the 10 tests, it shows that the sensor has a good distance reading capability with a very low error.

3.3. Sensor Load Cell Testing

This test is done by spilling oil (diesel fuel) on the water in the aquarium. Oil will be spilled on the surface of the water with a certain thickness then ADC data will be recorded. Figure 18 shows the TEMT6000 sensor testing to retrieve ADC data from the light intensity emitted by the LED. The following Table 4.5 is the result of ADC data from the taking made on the TEMT6000 sensor.

Figure 8. Sensor Load Cell Test.

Table 2. Sensor load cell testing data

Testing	Load(gr)	Sensor load cell(gr)	Error (%)
1	50.00	51.67	3.34 %
2	100.00	100.50	0.5 %
3	150.00	149.00	0.66 %
4	200.00	199.00	0.5 %
5	250.00	246.33	1.46 %
6	300.00	296.00	1.33%
7	350.00	345.00	1.42 %
8	400.00	394.50	1.37 %
9	450.00	440.33	2.14 %
10	500.00	488.16	2.36 %
	Error average		=1.5 %

This study used 2 load cell sensors to detect the weight that would be in the problem test on the towercopter. Picture 9 is a series of load cell sensors, HX711 modules, and STM. In table 2 is a test load cell sensor using a workpiece weighing 50-500 grams. Out of the 10 tests showed that the sensor had a good load-reading capability with a very low error of 1.5%.

3.4. ELM testing

Extreme *Learning Machine* (ELM) applied to arrange the speed of the motor which then being used to arrange the speed of motor 1, 2, and 3 when the burden is placed in the middle, in the right corner or in the left corner. The test has done to know about the accuracy of the method, then to compare datasheet result and training result to know about the errors. The error during the test of ELM can be counted using the formula as follow:

(3.1)	

Table 3. No-load testing

IOP Conf. Series: Materials Science and Engineering 1175 (2021) 012013 doi:10.1088/1757-899X/1175/1/012013

Uji	Input			Output (rpm)									
	Tinggi (cm)	Beban 1 (gr)	Beban 2 (gr)	Data asli			Data training			Error M1 %	Error M2 %	Error M3 %	
				M1	M2	M3	M1	M2	M3	0,1180	0,4350	0,8719	
1	15	0,00	0,00	1600	1600	1600	1601,9	1606,96	1613,95	0,1081	0,1616	0,1434	
2	20	0,00	0,00	1610	1610	1610	1608,26	1612,60	1612,30	0,0263	0,0917	0,0110	
3	25	0,00	0,00	1620	1620	1620	1620,43	1618,51	1620,17	0,1784	0,0198	0,1536	
4	30	0,00	0,00	1625	1625	1625	1627,9	1624,67	1627,49	0,1227	0,0039	0,0672	
5	35	0,00	0,00	1630	1630	1630	1632	1629,93	1631,09	0,0174	0,0463	0,1533	
6	40	0,00	0,00	1635	1635	1635	1634,72	1634,24	1632,49	0,1193	0,1109	0,3177	
7	45	0,00	0,00	1640	1640	1640	1638,04	1638,18	1634,79	0,0463	0,1234	0,2026	
8	50	0,00	0,00	1645	1645	1645	1644,24	1642,97	1641,66	0,1700	0,0528	0,1627	
9	55	0,00	0,00	1650	1650	1650	1652,8	1649,12	1652,68	0,2155	0,0591	0,3304	
10	60	0,00	0,00	1655	1655	1655	1658,57	1655,97	1660,46	0,1122%	0,1104	0,2413%	Error rata-rata

Table 3 is the test without burden, by comparing the original data: data from the tool and training data so that from that test can be gotten the data or the average error M1 0,1122%, M2 0,1104%, M3 0,2413%.

4. Conclusion

The conclusion is that towercopter can be controlled via android to display setpoint graphs, control start and stop, and display data from sensors. The ELM test was successful, namely when testing with a load of 100 grams, a low error was obtained with an average error of the three motors, namely 0,1122%, M2 0,1104%, M3 0,2413%.

IOP Conf. Series: Materials Science and Engineering 1175 (2021) 012013 doi:10.1088/1757-899X/1175/1/012013

5. References

[1] Andika I G C P 2018 Quadcopter obstacle avoidance dengan sensor inframerah untuk pemantauan bencana alam melalui udara. *Jurnal Pendidikan Teknologi dan Kejuruan* pp 71-72

[2] Dharmawan A R H 2012 Sistem stabilisasi posisi kamera pada quadcopter. *IJEIS* **110**

[3] Vonnete M M L A L 2018. Sistem pengukuran indeks massa tubuh menggunakan sensor jarak infra merah dan load cell. *Jurnal Fisika* **45**

[4] Prakoso T B 2015 Implementasi dji naza M-lite pada quadcopter. *Jurnal Ilmiah Mikrotek* **194**

[5] Fardani D P E W 2015 Sistem pendukung keputusan peramalan jumlah kunjungan pasien menggunakan metode extreme learning machine (Studi kasus: Poli gigi RSU DR.Wahidin Sudiro Husodo Mojokerto). *Journal of Information Systems Engineering and Business Intelligence* **34**

[6] Mardiana D 2018 Rancang bangun pengendalian sistem ballast kapal menggunakan metode ELM. (Surabaya: Politeknik Perkapalan Negeri Surabaya)

[7] Humaini Q 2015 Jaringan syaraf tiruan extreme learning machine (ELM) untuk memprediksi kondisi cuaca di wilayah Malang. (Malang: Fakultas Sains dan Teknologi Universitas Islam Negeri Maulana Malik Ibrahim Malang)

[8] Jami'in M A, Hu J, Marhaban M H, Sutrisno I and Mariun N B 2016 Quasi-ARX neural network based adaptive predictive control for nonlinear systems. *IEEJ Trans Elec Electron Eng* **11** pp 83-90

[9] Jami'in M A, Sutrisno I and Hu J 2012 Lyapunov learning algorithm for Quasi-ARX neural network to identification of nonlinear dynamical system. *IEEE International Conference on Systems, Man, and Cybernetics (SMC)* (Seoul, 2012) pp 3147-3152 doi: 10.1109/ICSMC.2012.6378275

[10] Jami'in M A, Sutrisno I and Hu J 2013 Deep searching for parameter estimation of the linear time invariant (LTI) system by using Quasi-ARX neural network. *The 2013 International Joint Conference on Neural Networks (IJCNN)* (Dallas, TX: 2013) pp 1-5 doi: 10.1109/IJCNN.2013.6707091

[11] Jami'in M A, Sutrisno I and Hu J 2015 Maximum power tracking control for a wind energy conversion system based on a quasi-ARX neural network model. *IEEJ Trans Elec Electron Eng* **10** pp 368-375

[12] Jami'in M A, Sutrisno I, Hu J, Mariun N B and Marhaban M H 2014 An adaptive predictive control based on a quasi-ARX neural network model. *13th International Conference on Control Automation Robotics & Vision (ICARCV)* (Singapore, 2014) pp 253-258 doi: 10.1109/ICARCV.2014.7064314

[13] Sutrisno I, Jami'in M A and Hu J 2014 Modified fuzzy adaptive controller applied to nonlinear systems modeled under quasi-ARX neural network. *Artif Life Robotics* **19** pp 22–26. https://doi.org/10.1007/s10015-013-0137-6

[14] Sutrisno I, Jami'in M A and Hu J 2014 An improved elman neural network controller based on quasi-ARX neural network for nonlinear systems. *IEEJ Trans Elec Electron Eng* **9** pp 494-501

[15] Sutrisno I, Che C and Hu J 2014 An improved adaptive switching control based on quasi-ARX neural network for nonlinear systems. *Artif Life Robotics* **19** pp 347–353

[16] Sutrisno I, Jami'in M A and Hu J 2012 Neural predictive controller of nonlinear systems based on quasi-ARX neural network. *18th International Conference on Automation and Computing (ICAC)* (Loughborough, 2012) pp 1-6

[17] Rinanto N and Khumaidi *Aplikasi Android-Rasspberry Pi pada Kapal Tanpa Awak untuk Pencarian Korban Kecelakaan Laut*

[18] Widodo R B, Quita R M, Amrizal S, Gunawan R S, Wada C, Ardiansyah A, Sutrisno I, Rahmat M B, Arfianto A Z, Handoko C R, Santoso A D, Santosa A W B, Ardhana V Y P and

Setiawan E Grasping and attached mode in human computer interaction in the study of mouse substitution. *Journal of Physics: Conference Series* **1196** 1 012034

[19] Sutrisno I, Jami'in M A, Hu J and Marhaban M H 2015 Self-organizing quasi-linear ARX RBFN modeling for identification and control of nonlinear systems. *54th Annual Conference of The Society of Instrument and Control Engineers of Japan (SICE)*. IEEE

[20] Jami'in M A, Sutrisno I and Hu J 2014 Nonlinear adaptive control for wind energy conversion systems based on quasi-ARX neural networksmodel. *Procedings of the International MultiConference of Engineers and Computer Scientists*. March 12-14, 2014 Hong Kong

[21] Sutrisno I 2009 Pemrograman komputer dengan software Matlab. (Surabaya: ITS Press)

IOP Conf. Series: Materials Science and Engineering 1175 (2021) 012014 doi:10.1088/1757-899X/1175/1/012014

Experimental study of air velocity effect of heat transfer in distillation condenser

AWibolo[1], A A I Made[2], Sudirman[3]
[1,2,3]Bali State Polytechnic Badung,Indonesia

Email: wibolo@pnb.ac.id

Abstract. A cooling tower is a heat exchanger equipment that uses water splashing media to lower the temperature by extracting heat from the water and emitting it into the atmosphere. Wind speed or velocity of wind waves is a fundamental atmospheric quantity caused by moving air from high pressure to low pressure while wind is moving air. Air velocity has an effect on the value of heat transfer coefficient, the faster the air, the better the value so that the heat transfer that occurs is also better. Research was also carried out on the human body to see forced convection due to the movement of these parts of the body, resulting in that the faster the body moves, the better the heat transfer coefficient value, this is related to the air flow that occurs on the body surface. The test shows that the more blower rotation make faster the heat transfer that occurs. At 2500 RPM rotation shows higher heat transfer by producing an average of 29.210 watts, the displacement decreases when the rotation is lowered to 1700 RPM to produce an average of 20.677 watts. The effectiveness of the tool by increasing the rotation from 1700 to 2500 by 32%.

1.Introduction

Heat Exchanger is a means of transferring/exchanging heat between fluid and other fluids through a separation wall. Based on the direction of fluid flow, this heat exchanger is a Cross Flow type (transverse / perpendicular), then the hot air that will enter the drying room will be separated from the smoke produced by the fuel [1]. Hot air is flowed in the arrangement of pipes (Tube Sheet) and sucked by a blower which is then forwarded into the drying room, while fuel combustion is carried out under the arrangement of pipes so that hot air with smoke from the results of separate combustion [2]. A heat exchanger is a device used to exchange energy in the form of hot water fluid of different temperatures which can occur through direct or indirect contact. The heat exchanger functions to release heat into the surrounding air which is located at the back of the old model refrigeration and is an arrangement of wires and vessels [3].

A cooling tower is one of the heat exchanger equipment that uses water splashing media to lower the temperature by extracting heat from water and emitting it into the atmosphere. Cooling tower is a heat exchanger that cools water through the air by being sprayed which causes a small portion of the water to evaporate. Cooling towers use evaporation where some of the water is evaporated into a moving air stream and then discharged into the atmosphere. As a result, the remaining water is significantly cooled. Cooling towers can lower the water temperature more than equipment that uses only air to dissipate heat, such as radiators in cars, and is therefore more cost effective and energy efficient. The cooling tower has several main parts including a fan or blower that functions to circulate air.

Content from this work may be used under the terms of the Creative Commons Attribution 3.0 licence. Any further distribution of this work must maintain attribution to the author(s) and the title of the work, journal citation and DOI.
Published under licence by IOP Publishing Ltd

Wind speed, or wind wave velocity, is a fundamental atmospheric quantity. Wind speed is caused by moving air from high pressure to low pressure, usually due to temperature changes, whereas wind is moving air (Wikipedia). The heat transfer rate increases with the increase in air speed. The flow of air outside the heat channel walls will cause forced convection [4]. Irvan Paramananda's research in 2014[5] by varying the air velocity on cooling with a radiator found that air velocity affects the effectiveness of heat transfer, the faster the air, the faster the heat transfer rate will be. Air velocity also has an effect on the value of heat transfer coefficient, the faster the air, the better the value so that the heat transfer that occurs is also better[6]. Research has also been carried out on the human body to look at forced convection due to the movement of these parts of the body. This research shows that the faster the body moves, the better the heat transfer coefficient value, this is related to the air flow that occurs on the body surface[7]. Some of these studies have stated that air velocity can affect the heat transfer that occurs.

This research will conduct an experiment on these results on a condenser cooling tower model distillation. This condenser has been made in previous research to produce good heat transfer due to changes in the cooling water flow rate [8]. Further research to further improve the effectiveness of the tool is to vary the blower air speed. The aim is to speed up heat transfer so that the phase change process becomes faster and will speed up the overall distillation time.

2.Research Methods

2.1.Design

Making tools has been done in previous studies, but redesign or shape changes are needed to readjust the tools to be installed according to the needs of the research being carried out. Several additional measuring instruments will also be installed to determine the phenomenon of temperature change data so that the heat transfer that occurs can be known significantly.

To determine the effectiveness of the cooling tower based distillation tube device, a test was carried out in the form of an analysis of the effect of air velocity on heat transfer so that it can accelerate the phase change. This test uses relevant theories and real data collection and refers to fixed variables that have been determined.

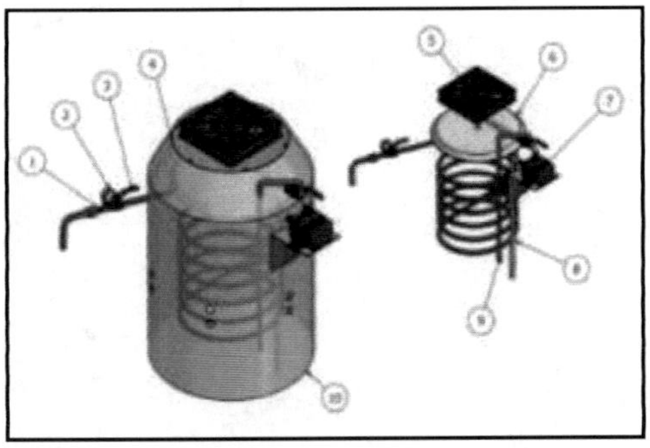

Figure 1. Distillation tool

1.Cooling pipe	6.Shower
2.Pressure gauge.	7.Water pump
3.Valve	8.Water pipe
4.Tube cover	9.Bypass pipe
5.Exhoust fab	10.Tube condenser

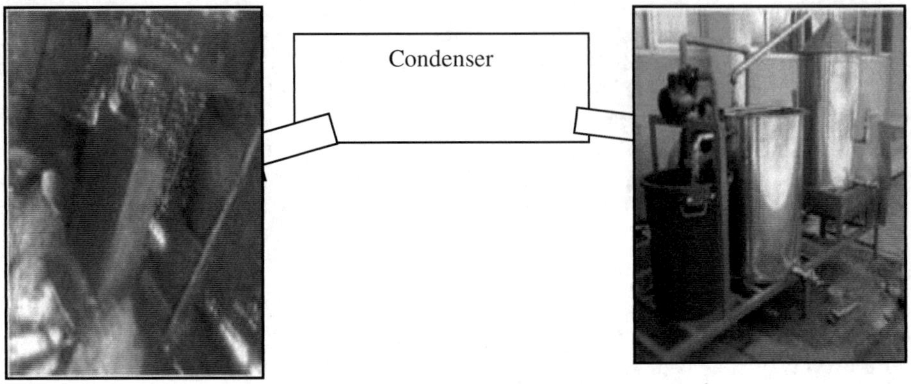

Figure 2. Condenser design

2.2.Research instrument

Tests are carried out based on the data that has been taken and fixed variables such as time and water discharge used. The test will be limited to only 5 attempts.

Measurements are made at several points to determine the heat transfer that occurs. The placement of the measuring instrument is carried out at several points to measure temperature. The measuring instrument sensor is placed at several points in accordance with Figure 3. The placement of the sensor is as needed. The T4 sensor inside the pipe is in a floating position to measure the temperature of the steam produced. The T3 sensor inside the pipe is attached to the wall to measure the temperature. The heat transfer from T4 to T3 is convection heat. The T2 sensor is placed outside against the pipe wall. The heat transfer from T3 to T2 is conduction heat. The sensor T1 is placed floating to measure the condenser room temperature. The heat transfer from T2 to T1 is convection heat. The total heat transfer is the sum of the three processes.

The data collection process is carried out by following the prepared flow. The reactor which has been filled with distillation material is heated according to the required temperature. After the temperature is reached (90 degrees C) the steam valve is opened. The test is carried out by varying the fan rotation to regulate the amount of air sucked from the condenser. Data were collected every 10 minutes in a span of 90 minutes.

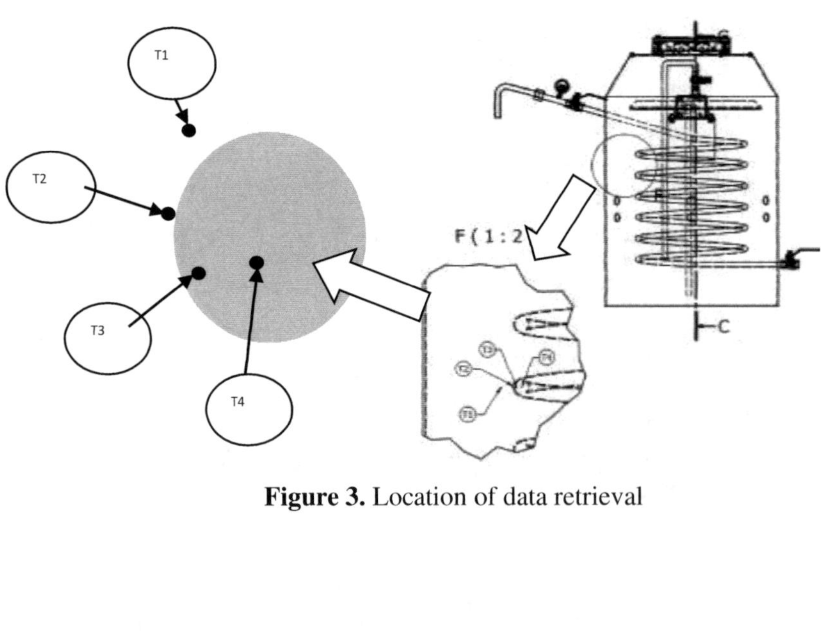

Figure 3. Location of data retrieval

3.Results

3.1.Result Of Data

The test is done by varying the blower rotation from low RPM to high RPM. Change using a potentiometer and rotation measured with a tachometer. The higher the RPM, the greater the volume of air flowing. The rotation is adjusted from RPM 1700, 1900, 2100, 2300 and 2500. Test result data are as in table 1.

Table 1. Test Result Data

No		Heat Transfer (Q)				
	RPM	2500	2300	2100	1900	1700
1		26.943	26.943	23.865	23.865	18.013
2		29.869	26.943	24.017	20.939	18.013
3		27.095	RPM 02	20.939	18.013	17.861
4		26.639	23.865	21.091	19.628	18.165
5		26.803	26.791	23.865	23.865	20.939
6		31.040	26.943	26.943	23.865	23.865
7		32.643	26.791	23.865	25.480	20.939
8		31.332	26.791	24.017	25.480	20.939
9		29.869	28.406	26.943	26.943	24.017
10		29.869	28.406	26.943	26.943	24.017
Rate of Q		29.210	26.428	24.249	23.502	20.677

Figure 4. Heat Transfer Graph

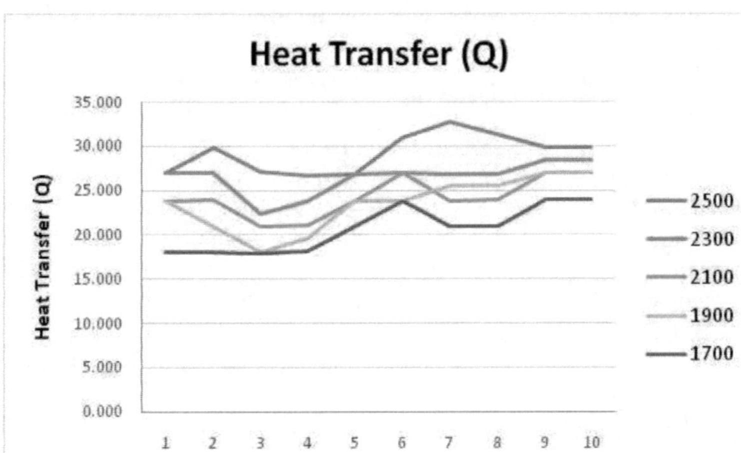

3.2.Descriptive Analysis

The test data shows that the higher the blower rotation, the faster the heat transfer occurs. At 2500 RPM rotation shows a higher heat transfer by producing an average of 29.210 watts, the displacement decreases when the rotation is lowered to 1700 RPM to produce an average of 20.677 watts.

This phenomenon occurs because the volume of flowing air increases. The air is flowed through the hole on the bottom side of the condenser through the steam pipe and then discharged at the top of the condenser. The air that flows uses room air where the condenser is placed at a temperature of 25 degrees C. The air is mixed with water which is a splash from the top of the condenser. Water uses a circulating pump. Water is sucked from the bottom of the condenser and then flowed upwards to be

IOP Conf. Series: Materials Science and Engineering 1175 (2021) 012014 doi:10.1088/1757-899X/1175/1/012014

sprayed in the form of a splash. The water discharge is kept constant. The combination of flowing air and water gives a nice cooling effect to the designed condenser.

The effectiveness of the tool by increasing the rotation from 1700 to 2500 is 32%. The effectiveness states that the higher the blower rotation, the better the heat transfer.

Future, research will focus on the pressure in the condenser tube. The existing theory states that pressure give an affects the heat transfer. Changing the pressure is done by closing the air inlet and varying the fan speed. This can change the air pressure in the tube. Hopefully, the lower of the air pressure, the better the heat transfer that occurs. By increase of heat transfer will affect the rate of vapor phase change in the condenser pipe. This effort was made to obtain an effective condenser design and reduce energy use.

4. Conclusion

This research has resulted in changes that occur due to varying the rotation of the air fan. Expectations from this research have been derived from these changes. Several things need to be done again to increase the effectiveness of the condenser designed.

This research can give the following conclusions are :

1. The increase in blower rotation affects the heat transfer that occurs in the cooling tower model distillation condenser.
2. The increase in rotation can increase the effectiveness of the cooling tower model distillation condenser.

5. References

[1] Satish Kumar M V H 2016 Performance analysis of cooling tower. *International Journal of Engineering Trends and Technology (IJETT)* **38** 9

[2] Sulaeman 2014 Effectiveness analysis of heat exchanger. *Journal of Machine Technique Padang Technology Institute* **4** 1

[3] Putra T D 2014 Instrumentation system planning at the design of test equipment for wire and tube heat exchanger efficiency. *JRM Technique Faculty Surabaya University* **1** 2

[4] Chaibi M T 2013 Experimental analysis of the performance of a mechanical geothermal water cooling tower in South Tunisia. *American Journal of Energy Research- Institut National de Recherche en Génie Rural Eauxet Forets* **1** 1 (Ariana: Tunisia)

[5] Irvan Paramananda and Prabowo Prabowo 2014 Studi Eksperimen Pengaruh Variasi Kecepatan Udara Terhadap Performa Heat Exchanger Jenis Compact Heat Exchanger (Radiator) Dengan Susunan Tube Inline Sebagai Pemanas Pada Sistem Pengeringan Batubara. *Jurnal Tehnik ITS* **3** B-91

[6] Xiao, B., Wang, G., Wang, Q., Maniruzzaman, M., Sisson, R.D. and Rong, Y., 2011. An experimental study of heat transfer during forced air convection. *Journal of materials engineering and performance* **20** 7

[7] Oliveira, A.V.M., Gaspar, A.R. and Quintela, D.A., 2011. Dynamic clothing insulation. Measurements with a thermal manikin operating under the thermal comfort regulation mode. *Applied Ergonomics* **42** 6

[8] Adiaksa, I.M.A. and Wibolo, A., 2020, February. Experimental analysis of fluids discharge in cooling tower condenser distillation model. In *Journal of Physics: Conference Series* **1450** 1

IOP Conf. Series: Materials Science and Engineering 1175 (2021) 012015 doi:10.1088/1757-899X/1175/1/012015

Assesment dermoscopy images of skin lesion using U-Net segmentation for clinicians teledermatology

A Ciptaningrum[1], I K E Purnama[2], R F Rachmadi[3]
[1]Department of Electrical Engineering
[2]Department of Electrical Engineering and Department of Computer Engineering
[3]Department of Computer Engineering
Institut Teknologi Sepuluh Nopember Surabaya

E-mail: adiratna.18071@mhs.its.ac.id

Abstract. Standard solutions to the process of assessing skin lesions are based on a medical pathology examination using dermoscopy images. Fortunately, abovementioned-mentioned diagnosis and treatment quite often show a defect-prone, yet in the hands of accomplished health care professionals. Due to the rapid development of skin cancer lesions, computational analysis is required. In order to minimize the probability of error, a segmentation task is required in which multiple medical analyzes can be conducted. These frameworks are generally required versatile and need assistance from advanced computing power. The proposed system is tested with datasets for dermoscopy images of clinical signs. U-Net segmentation method provides greater segmentation result IoU 94.37, Dice Coefficient 88.11, precision 90.87, recall (sensitivity) 91.82, accuracy 94.55, loss 16.8, and F1-score 91.34.

1. Introduction

Skin cancer is known as irregular proliferation of skin cells, most commonly occurs on sun-exposed skin, and is the most prevalent type of malignancy in humans. The most dangerous skin cancers are melanoma, squamous cell carcinoma, and basal cell carcinoma. Malignant melanoma is identified as melanoma, the most severe form of skin cancer [1]. Malignant melanoma arises in melanocytes, where the cells produce melanin and add color to the skin. This case has a rising trend in the world, particularly for women under the age of 40. However, this recently retired cancer will be treated if it is detected early. As we know, doctors inspect and capture their patients with visual examination assisted by hand-held dermoscopy. The possibility visual analysis would catch both the macroscopic and the microscopic. The macroscope captures the optical close-up, and the microscope is also known as the dermoscopy [2].

Dermoscopy is a way to examine the skin using bright light and enhancement, and also using polarizing or immersion fluids to minimize surface reflections. Dermoscopy has only improved the rate of diagnosis by visual examination. The dermoscopy images image plays an important role in decision-making to help the diagnosis of a disease. This is attributed to the fact that the image offers different details found in it for study. Quantitative analysis of digital images can yield high throughput images, such as image segmentation tasks. This segmentation approach is the most preferred condition for automatic image processing, and most researchers define and analyze each lesion in the study to help clinicians measure how they respond to different treatments [3].

Recently, the image segmentation task is a crucial step needed to establish the scale and ratio that is used for clinically relevant purposes. As we know, the production of automated segmentation is difficult due to wide differences in texture, shape and colour. Various types of image processing

Content from this work may be used under the terms of the Creative Commons Attribution 3.0 licence. Any further distribution of this work must maintain attribution to the author(s) and the title of the work, journal citation and DOI.
Published under licence by IOP Publishing Ltd

methods are provided, such as regional growth, thresholding, morphological operations, k-means clustering, and most of their accuracy are less common types. In fact, the noise and background of the image make it impossible to achieve thorough and accurate segmentation. In addition, artifacts may be produced during image processing and the contrast is weak [3].Actually, the most preferred approach is to take a deep-learning section of the lesion. One of the learning architectures for the segmentation of images is the U-Net. In general, this approach uses a downsampling encoder and an upsampling decoder architecture by separating multiple parameters for each level. This architecture has the potential to run at a much greater random sample without requiring any extra effort. U-Net is also suitable for sample processing and automated large-scale recording. U-Net is more competitive than current methodologies in terms of architecture and pixel-based image segmentation generated by convolutional neural network layers. It is also efficient with small dataset images. The value of this network system is that it can not only effectively segment the targeted lesion target and efficiently analyze and objectively examine the dermoscopy image of the skin lesion, but also help to increase the accuracy of medical image assessment. This study consequently provides a systematic literature review of U-net-based medical image segmentation, focusing on the effective segmentation implementation of U-net for medical imaging systems.

2. Related Works

On that same basic principle of a research study, a deep learning algorithm is the most preferred method of image segmentation to solve this issue. For such segment lesion region, the features are systematically trained in deep learning because they achieve significant output of the metrics, segment the region, and require a huge number of annotated data. The Convolutional Neural Network, known as CNN, has been able and efficient in large amounts of data. Previous studies by C. Hernandez et.al. [4] Use Function Pyramid Network (FPN) and merge neural nets with VGG-style to predict the cell mask. F. Araújo et.al, meanwhile. [5] used CNN to estimate the irregular region of cells in the pap test.

In addition, Bi et al.[6] suggested segmentation of skin lesions using multi-stage, completely convoluted networks (FCNs). The multi-stage method involved localized early-stage coarse-appearance learning and also detailed boundary-characteristic learning in the next stage. In his work, a parallel integration strategy was put in order to make a merger of the outcome that they reported to have improved detection. However, various PH2 datasets outperformed their system by around 90.66% and still show modest change. T. Tran et al. [7] use of these Seg-Net in white blood cells (leukocytes) and red blood cells (erythrocytes) in peripheral blood tests shows a global accuracy of 89.45%. Between multi cell or cell nuclei segmentation, U-Net is the most recommended and always provides state-of-the-art segmentation outcomes.

In comparison, R. Hollandi et.al. [8] used two approaches in the form of R-CNN and U-Net to estimate cell nuclei, and the algorithm proposed to measure the level of accuracy. And then, in the analysis of Pan et.al[9], simplified reconstruction techniques based on CNN were used, the nuclei of which is roughly segmented from the background. In Yang et al.[10] 's work, Fully Convolutional Neural Network (F-CNN) was used for an iterative k-terminal trying to cut algorithm for the segmentation of glia cells. The F-CNN has used a fully connected layer with a convolution layer at the end, even though other layers extract the necessary features from input data, so this method is best suited. Akram et al.[11] demonstrated a concept for a cell bounding box and used F-CNN for cell segmentation to achieve good cell segmentation accuracy based on spatial details. Hou et al.[12] applied CNN to the nuclei section using the Generative Adversarial Network (GAN) module to generate synthetic patches dependent on weight. However, segmentation collapsed when the nuclei is close to the dark zone. Hatipoglu et al.[13] used the spatial and temporal connections of extracellular and cell pixels. It can segment cells by generating a number of pixels in the window framework to fit neighboring pixels. U-Net is the most favored approach since there has been a major change in the segmentation and classification of the medical image.

There are three reasons U-Net is used for this work. First, the ambiguity of its boundaries makes it very hard to differentiate the gland from surrounding tissue with intraprostatic tissue heterogeneity further contributing to under or oversegmentation. Second, examinations on different dermoscopic skin

IOP Conf. Series: Materials Science and Engineering 1175 (2021) 012015 doi:10.1088/1757-899X/1175/1/012015

lesion with use of different imaging protocols lead to wide variations in signal intensity. Third, the skin lesion has a wide variety of sizes, textures of tissue, either due to anatomical differences within patients or due to the presence of pathology.

3.Material and Methods

3.1. Dataset

The data of dermoscopic skin lesion is provided by PH2 dataset [14]. There are three unqiue diagnose of skin lesion of dermoscopic such as melanoma, nevus, and seborrheic keratosis. Melanoma is malignant skin tumor that derived from melanocytes (melanocytic). It can be seen in Figure 1.

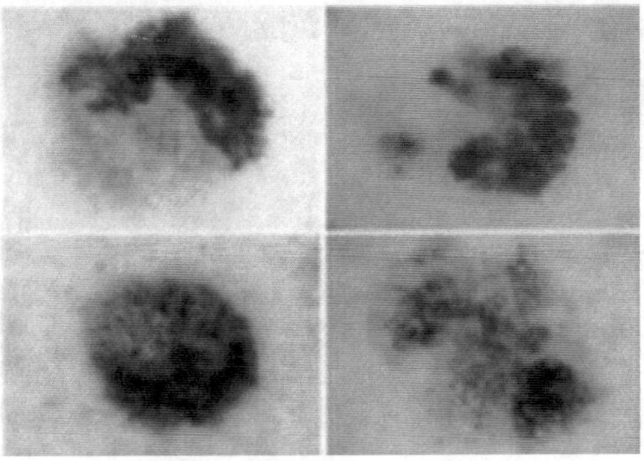

Figure 1. Melanoma.

Nevus is benign skin tumor that derived from melanocytes (melanocytic) can be seen in Figure 2. And then, seborrheic keratosis is benign skin tumor, derived from keratinocytes (non-melanocytic) can be seen in Figure 3.

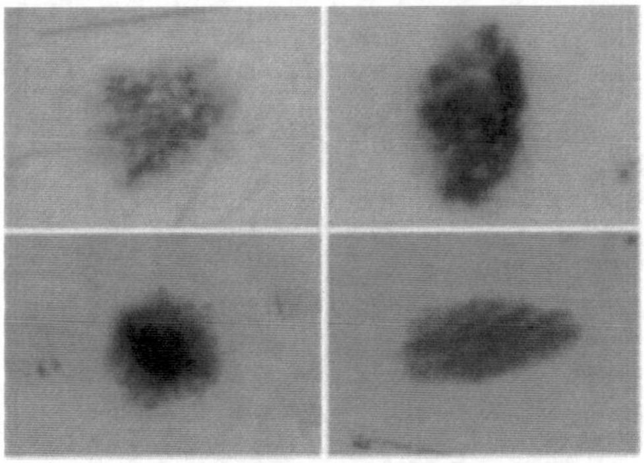

Figure 2. Nevus.

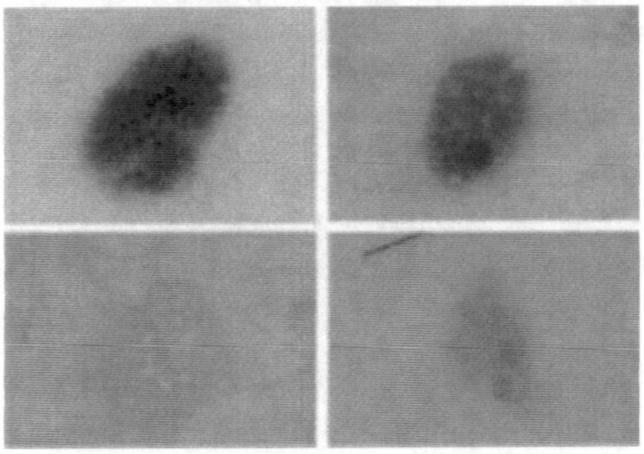

Figure 3. Seborrheic keratosis.

IOP Conf. Series: Materials Science and Engineering 1175 (2021) 012015 doi:10.1088/1757-899X/1175/1/012015

Lesion information contains the original image, combined with a gold standard (definitive) diagnosis, which have seven groups such as melanocyte nevi, melanoma, benign keratosis, basal cell carcinoma, actinic keratoses, vascular lesions, dermatofibrima, and so-called ground truth. This dataset includes 200 images as training data of which 160 images are naevus (atypical naevus and normal naevus), as well as 41 images of melanoma, and each lesion is followed by a binary mask identified by an expert. These images resized as 256×256 to maintain training. U-Nets were chosen due to their proven ability to segment microscopic images.

3.2. The Structural Of U-Net Models

U-Net is also known as the original semantic segmentation network, who have generated incredible results in the segmentation of medical images. This architecture can be expansively thought of as an encoder network followed by a decoder network. By integrating layers, the encoder bit by bit eliminates the spatial dimension. And then bit by bit the decoder recovers the information of the object and the spatial dimension. Besides that, there are also fast connections, where the decoder encoder allows the decoder to better recover the information of the object. As a highly invariant, the U-Net provides a stable and precise segmentation of its form, where the features of the images are combined to obtain high-resolution data. In addition, the pooling operator is replaced by an upsampling operator to increase the output resolution. Adding more layers will help remove additional data input features and improve significantly accuracy. The architecture of the U-net architecture can be seen in Figure 1.

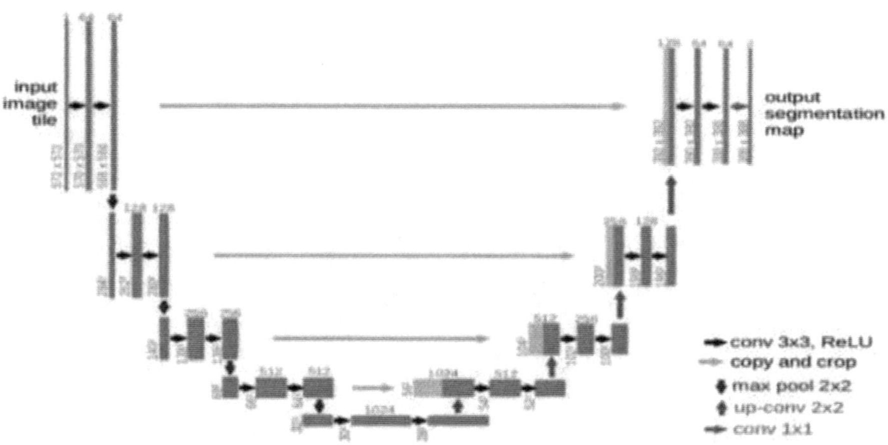

Figure 4. U-Net Architecture.

The breakdown of U-Net architecture can be seen in Figure 4. Each mechanism consists of two convolutional layer, and the number of channels varies from 1 to 64.

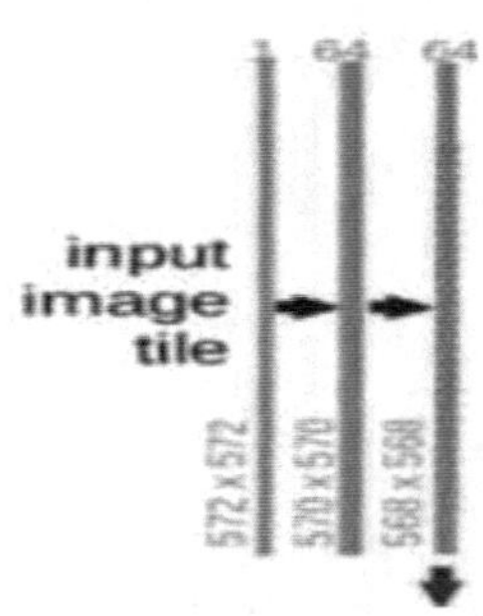

Figure 5. Input Image.

IOP Conf. Series: Materials Science and Engineering 1175 (2021) 012015 doi:10.1088/1757-899X/1175/1/012015

As a convolutional process, the depth of the image will increase. The red arrow pointing downward is the max pooling process which is half the downsizing of the image. The downsizing is the size reduced from $572 \times 572 \rightarrow 568 \times 568$ is due to padding, but the implementation still using the same padding.

Figure 6. The process is repeated 3 more times.

In Figure 6 can be seen that U-Net reaches at the bottom. Still two convolutional layers are built, but with no max pooling. The image at this bottom has been resized to 28×28 1024. The next step is expansive path, can be seen in Figure 7.

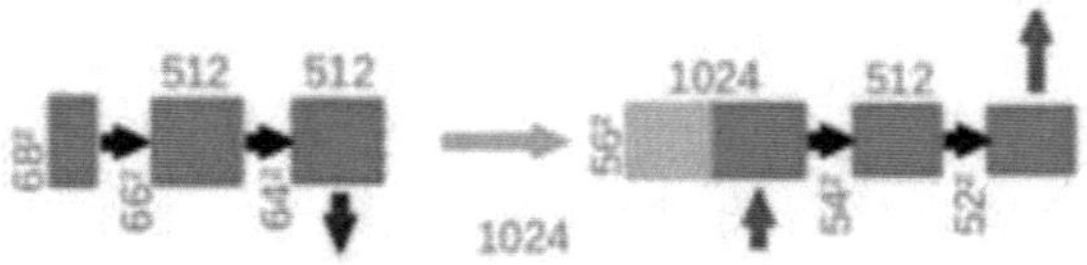

Figure 7. The process is repeated 3 more times.

In the expansive path, the image is upsized to its original size, and need transpose convolution. This transpose convolution is an upsampling technic that expands the size of images. Basically, the transpose convolution does some padding on the original image followed by a convolution operation. After the transposed convolution, the image is upsized from $28 \times 28 \times 1024$ to $56 \times 56 \times 512$. In this stage, the image is merged with the appropriate image from the contracting line and makes an image in size $56 \times 56 \times 1024$ to combine the information from the previous layers in order to get a more precise prediction.

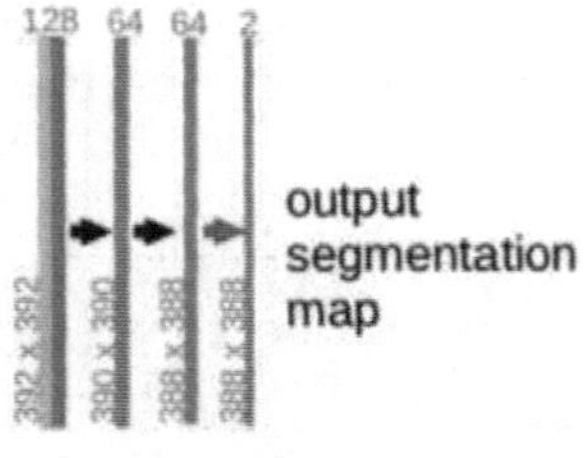

Figure 8. The process is repeated 3 more times.

The uppermost of the architecture is the last step to reshape the image to make prediction requirements. The last layer is a convolution layer with one filter of size 1×1. And then, the rest left one is the same for neural network training. During training are used deep models based on the popular U-Net architecture. Adam optimization was chosen to update the weights with binary cross entropy as

IOP Conf. Series: Materials Science and Engineering 1175 (2021) 012015 doi:10.1088/1757-899X/1175/1/012015

loss function. In this method, loss function is used combination of binary cross-entropy and binary Dice loss to train the model. The Dice loss is to make segmented lesion, so it will equal to ground truth.

3.3. Loss Function

In semantic segmentation image is actually a pixel-level classification. Dice loss is to make segmenting the target of skin lesion to be equal to each other with ground truth as much as possible. The Eq. (1) of Dice coefficient is :

$$Dice = 2 \times \frac{y_a \cap y_b}{|y_a| + |y_b|} \quad \text{where} \quad dice \in (0,1) \tag{1}$$

In order to maximize the penalty, the lower the estimation result of the Dice coefficient is done by logarithmic (log) operations and negative values are taken as a loss function, e.g. Log-Dice loss. Then in (2) is shown where y_a is the ground truth value of skin lesion image and y_b is the predicted value.

$$LogDL = -\log(2 \times (y_a \cap y_b))/(|y_a| + |y_b|) \tag{2}$$

In addition, Focal Loss may inhibit the contribution of easily segmented pixel regions to model loss, thereby enhancing regions that are not easily segmented. The theorem can therefore be shown in (2) as below.

$$Focal\ Loss = y_a \times \log(y_b) \times (1 - y_b)^{\lambda} \tag{3}$$

Consider the two equations, Log-Dice loss and Focal loss are combined as a loss function for model training by parameter λ. The λ specifies the proportion of Log-Dice loss and Focal loss in the overall loss function. The optimization of λ can also be chosen as corresponding to the values of F1-score and Intersection over Union (IoU). It can be seen in (3).

$$Loss = \lambda \times \log DL + (1 - \lambda) \times Focal\ Loss \tag{4}$$

4.Result and Discussion

The proposed framework of U-Net to efficiently extract relevant features in the training process. This architecture clearly outperforms traditional graphic-based segmentation method. In this on experimental is validated and testing using PH2 dataset, which have 201 images with its dimension 256×256. However, the new transformed images that will be added after image augmentation process.

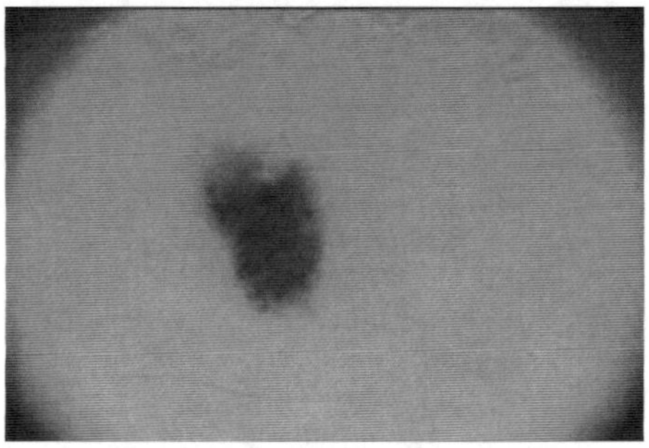

Figure 9. Original Image.

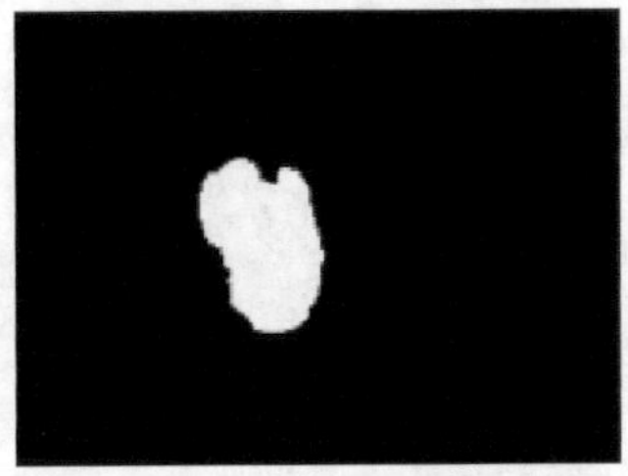

Figure 10. Corresponding Mask.

The performance of this segmentation is evaluated using performance metrics. The metrics based on accuracy, precision, recall, F1-score, and IoU which has a representation of each. In this work, the accuracy means the ratio of the accurately segmented the lesion and background pixels to all the pixels in the image. Between the total of segmented skin lesion pixels, the precision represent of segmented skin lesion in label images, and it is called positive predictive value. Later, recall defines the percentage of the total numbers of lesion images on the label segmented by the proposed technique or called sensitivity. The F1-score is used to calculate the average accuracy and recall. And the last one, IoU, is the correlation between the predicted values and the ground values of truth. In the Eq. (5) (7) (8) (9), TP, FP, FN and TN are true positive , false positive , false negative and true negative.

$$Accuracy = \frac{TP+TN}{TP+FP+FN+TN} \tag{5}$$

$$Precision = \frac{TP}{TP+FP} \tag{6}$$

$$Recall = \frac{TP}{TP+FN} \tag{7}$$

$$F1 - Score = \frac{2 \times Precision \times Recall}{Precision+Recall} \tag{8}$$

$$IoU = \frac{|y_a \cap y_b|}{|y_a| + |y_b| - |y_a \cap y_b|} \tag{9}$$

In Eq. (4) First, the λ parameter of loss function is selected. And then, the best optimization method and loss function are chosen. The second one, skin lesion segmentation in datasets are processed using the proposed semantic segmentation as the model. The highest output and training network is chosen to boost the algorithm. The next step, then, is to evaluate the efficiency of the enhanced model with the same dataset. In this experiment, the comparison of Log-Dice loss and Focal loss with Adam is modified as an optimization. A massive quantity of data is required to create an efficient information classifier. An enhancements in image is expected to improve the performance of this network. In this research, an enhancement is described, one for random rotation and one for horizontal flipping.

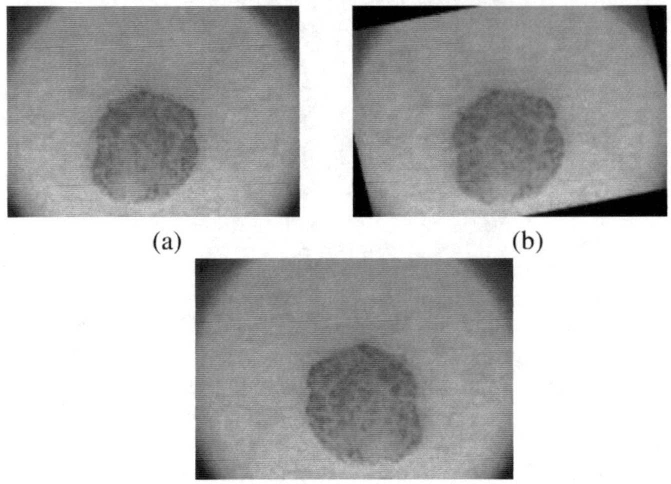

(a) (b) (c)

Figure 11. (a) Original Image (b) Rotated Image (c) Flipped Image

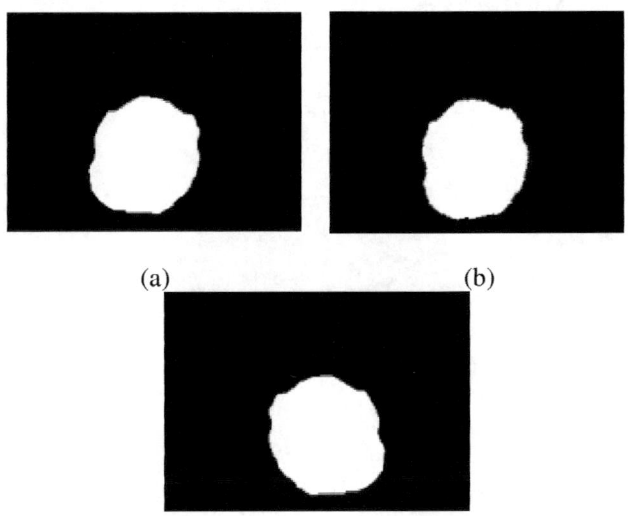

(a) (b) (c)

Figure 12. (a) Original Mask (b) Rotated Mask (c) Flipped Mask

This model performs well if just looking at the resultant images. In this model various of the epoch were tested in 125, 250, 750, and 1500.

Table 1. Experimental Performance On Skin Lesion Of U-Net 125 Epoch.

	Table Column 50 Epoch		
	Train Set	***Test Set***	***Vaidation. Set***
IoU	97.47	93.75	94.73
Dice Coefficient	86.58	80.87	82.19
Precision	96.85	89.58	92.00
Recall (Sensitivity)	97.52	91.31	92.14
Accuracy	98.22	94.03	94.93
Loss	10.25	18.49	16.55
F1-Score	97.18	90.44	92.03

Table 2. Experimental Performance On Skin Lesion Of U-Net 250 Epoch.

	Table Column 250 Epoch		
	Train Set	***Test Set***	***Validation Set***
IoU	98.28	94.27	95.12
Dice Coefficient	90.20	84.08	85.29
Precission	97.35	89.51	91.91
Recall (Sensitivity)	98.39	92.40	92.74
Accuracy	98.65	94.22	95.06
Loss	7.41	16.88	15.03
F1-Score	97.87	90.93	92.32

Table 3. Experimental Performance On Skin Lesion Of U-Net 750 Epoch.

	Table Column 750 Epoch		
	Train Set	***Test Set***	***Validation Set***
IoU	98.95	94.28	95.48
Dice Coefficient	93.78	86.86	88.43
Precission	98.40	91.12	92.94
Recall (Sensitivity)	98.77	91.04	92.49
Accuracy	99.11	94.48	95.34
Loss	4.65	16.73	14.00
F1-Score	98.59	91.08	92.71

Table 4. Experimental Performance On Skin Lesion Of U-Net 1500 Epoch.

	Table Column 1500 Epoch		
	Train Set	***Test Set***	***Vaidation. Set***
IoU	99.15	94.37	95.50
Dice Coefficient	95.00	88.11	89.52
Precision	98.80	90.87	93.11
Recall (Sensitivity)	98.82	91.82	92.60
Accuracy	99.25	94.55	95.43
Loss	3.74	16.80	13.90
F1-Score	98.81	91.34	92.85

IOP Conf. Series: Materials Science and Engineering 1175 (2021) 012015 doi:10.1088/1757-899X/1175/1/012015

As we can see in the Table 1, 2, 3 and 4, the bigger of epoch then more bigger Dice Coefficient result. So it represents the validated image segmentation performance has similar object size. In Figure 12, we compared the epoch of of train set, test set, and validation set on different epoch.

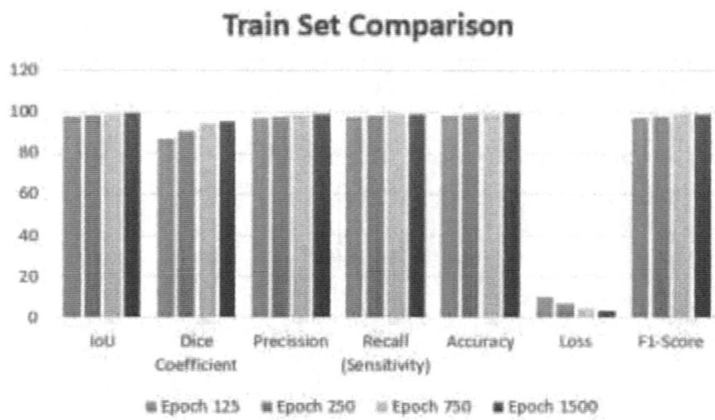

Figure 13. Comparison of metrics on train set.

In Figure 13. On train set we can see that the higher epoch then smaller the loss value whereas the Dice Coefficient and IoU more higher. However, it is the opposite of the test set where the higher the epoch value, the bigger the loss value. It can be seen in Figure 14.

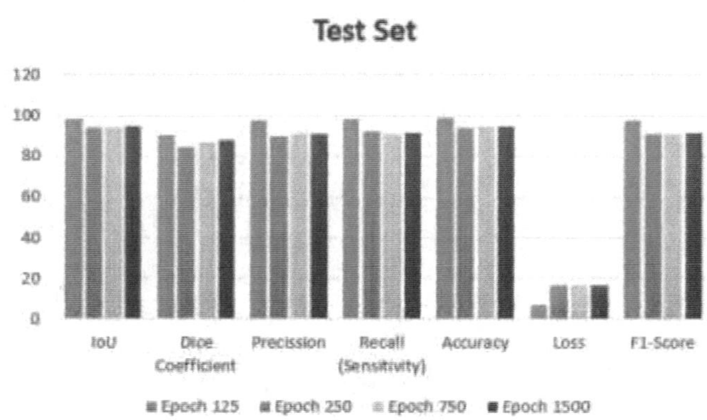

Figure 14. Comparison of metrics on train set.

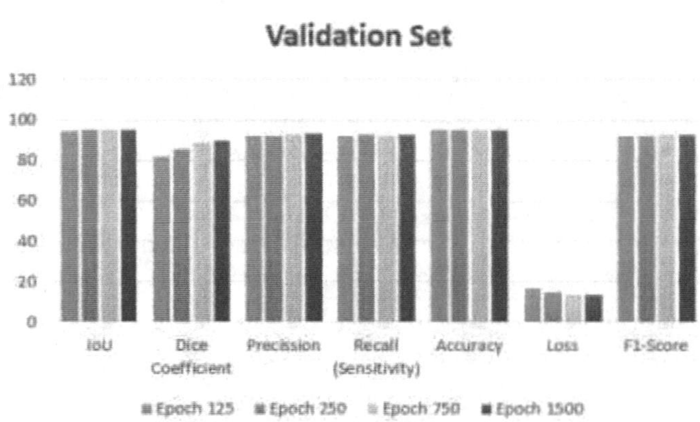

Figure 15. Comparison of metrics validation set.

After conducting model training in the training set, then make predictions in the unseen test set and visualize the predicted output, so it can be seen in Figure 17.

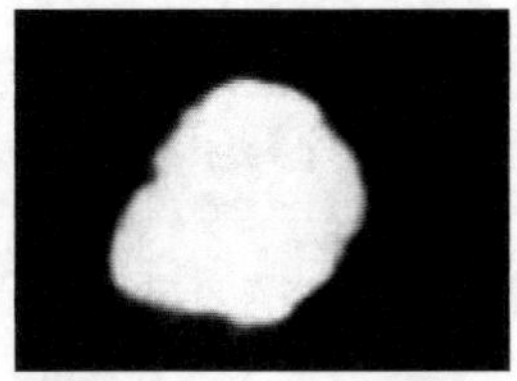

Figure 16. Visualize predicted lesion.

Currently, the predicted outputs are blurry enough because of the predicted pixel values are in the range 0 to 1. To make clear edge predictions, this work enhances the image by rounding the pixel value to $1 > 0.5$.

Figure 17 (a). Comparing the prediction after enhancement.

Figure 17 (b). Comparing the prediction after enhancement.

After visualing comparison of the prediction after enhanchement, this work applying the mask. The output can be seen in Figure 18.

Figure 18 (a). Output after applying mask: Predicted mask.

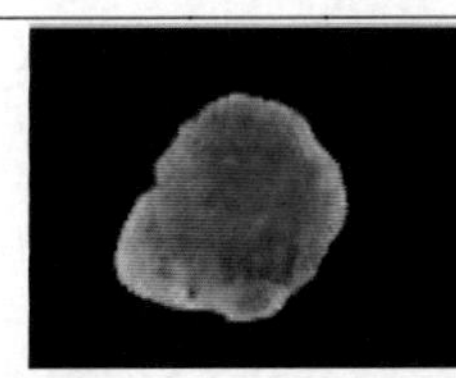

Figure 18 (b). Output after applying mask: Segmented output.

5.Conclusion

In this paper, we have proposed U-Net models as segmentation of dermoscopic skin lesion image. This experiment shows the segmentation results related to the proposed approach. In this experiment metrics metrics plays pivotal role during evaluating the performance such as IoU, Dice Coefficient, precission, recall (sensitivity), accuracy, and F1-score. The proposed architecture improves the performance of dermoscopic skin lesion segmentation in each area, such as visualization, training, testing, and validation.. This study only focuses for segmentation task using U-Net architecture on skin lesion dataset. Further improvement can be made by fine-tuning the hyper-parameters of U-Net. The future hope, U-Net algorithm can be tested on the other publicly available segmentation data medical and biology domains.

6. References

[1] Mermelstein R and Riesenberg L 1992 Changing knowledge and attitudes about skin cancer risk factors in adolescents. *Health Psychology* **11** 6 p 371

[2] Harvey N T and Wood B A 2018 A practical approach to the diagnosis of melanocytic lesions. *Arch Pathol Lab Med* **143** pp 789 - 810 doi: 10.5858/arpa.2017-0547-RA

[3] Errichetti E and Stinco G 2016 Dermoscopy in general dermatology: A practical overview. *Dermatol Ther* **6** pp 471 - 507 doi: 10.1007/s13555-016-0141-6

[4] Hernández C X, Sultan M M and Pande V S 2018 *Using Deep Learning for Segmentation and Counting within Microscopy Data*. arXiv:1802.10548 [cs, q-bio]

[5] Araújo F H D, Silva R R V, Ushizima D M, Rezende M T, Carneiro C M, Campos Bianchi A G and Medeiros F N S 2019 Deep learning for cell image segmentation and ranking. *Comput Med Imaging Graph* **72** pp 13 - 21

[6] Yu L, Chen H, Dou Q, Qin J and Heng P A 2017 Automated melanoma recognition in dermoscopy images via very deep residual networks. *IEEE Trans. Med. Imag* **36** 4 pp 994 – 1004, Apr 2017

[7] Tran T, Kwon O, Kwon K, Lee S and Kang K 2018 Blood cell images segmentation using deep learning semantic segmentation. *IEEE International Conference on Electronics and Communication Engineering (ICECE)* pp 13 - 16

[8] Hollandi Ret al 2019 *A Deep Learning Framework for Nucleus Segmentation using Image Style Transfer*. bioRxiv. 2019: 580605

[9] Pan X et al 2017 Accurate segmentation of nuclei in pathological images via sparse reconstruction and deep convolutional networks. *Neurocomputing* **229** pp 88 - 99

[10] Yang L, Zhang Y, Guldner I H., Zhang S and Chen D Z 2016 3d Segmentation of glial cells using fully convolutional networks and k-terminal cut. *Med Image Comput Comput Assist Interv of Lect Notes Comput Sci.* **9901** pp 658 - 666 (Springer)

[11] Akram S U, Kannala J, Eklund L and Heikkil¨a J 2016 Cell segmentation proposal network for microscopy image analysis. *DLMIA of Lect Notes Comput Sci.* **10008** pp 21 - 29

[12] Hou L et al 2019 Sparse autoencoder for unsupervised nucleus detection and representation in histopathology images. *Pattern Recognition* **86** pp 188 - 200

[13] Hatipoglu N and Bilgin G 2017 Cell segmentation in histopathological images with deep learning algorithms by utilizing spatial relationships. *Medical & Biological Engineering & Computing* **55**

[14] Mendonça T P M, Ferreira J S, Marques A R, Marcal and Rozeira J 2013 PH2 - A dermoscopic image database for research and benchmarking. *Proc. 35th Annu. Int. Conf. Eng. Med. Biol. Soc. (EMBC)* (July 2013) pp 5437 – 5440

IOP Conf. Series: Materials Science and Engineering 1175 (2021) 012016 doi:10.1088/1757-899X/1175/1/012016

Extraction of keywords in the abstract of scientifics papers using BERT

M Ikhsan[1], **R F Rachmadi**[2], **E M Yuniarno**[3]
[1]Department of Electrical Engineering
[2]Department of Electrical Engineering and Department of Computer Engineering
[3]Department of Computer Engineering
[1,2,3]Institut Teknologi Sepuluh Nopember Surabaya

E-mail: mikhsan.18071@mhs.its.ac.id

Abstract. Keywords are used to facilitate searches, inaccurate keyword determination, which can result in searches for journals that do not match the intended title. To overcome this, it is necessary to implement a keyword search in the abstract using deep learning methods from a journal to produce the right keywords that will distinguish journals from one another. The purpose of this study is to implement a keyword search system in abstracts using the Bidirectional Encoder Representations from Transformers (BERT) method as an abstract word search system that produces keyword answers that have context according to the desired keyword results. The results of this study get optimal results in the epoch 500, and from these results using a dataset with of 444 training data and 10 test data for testing.

1.Introduction

Keywords are an integral part of writing scientific articles. Keywords can be taken from the thesaurus according to their respective scientific fields, written in one line with a number varying from 3 to 6 words and sorted from specific to general. Keywords can inform the subject matter of a scientific article. In addition, keywords have an important role in article indexing. Article indexing can help make it easier to find an article through computer scanning on the internet, for example, if someone wants to search for an article, a keyword search can provide results in the form of articles that are relevant to the given keywords. Therefore, the selection of keywords must be done well, so that they can effectively describe the whole subject in a scientific article. In this era of machine learning, choosing effective keywords is possible and easy to do.

Machine learning has developed very rapidly and has been widely applied in various fields of research. One of them is in various text mining processes, such as word identification, word classification, entity extraction, word analysis, and other aspects of the text [1]. In recent years, there have been many studies that have developed text mining algorithms, one of which is the BERT (Bidirectional Encoder Representations from Transformers) algorithm [2]. BERT is a new technique in NLP (natural language processing) created by researchers from Google AI Language. In its creation, Google AI Language can present more modern results in a wide variety of problems in NLP, including Questionnaires, Natural Language Inference (MNLI), and others.

Based on the success of the research above, this study implements the BERT method in searching for abstracted keywords in scientific articles. The dataset used in this research is in the form of

Content from this work may be used under the terms of the Creative Commons Attribution 3.0 licence. Any further distribution of this work must maintain attribution to the author(s) and the title of the work, journal citation and DOI.
Published under licence by IOP Publishing Ltd

IOP Conf. Series: Materials Science and Engineering 1175 (2021) 012016 doi:10.1088/1757-899X/1175/1/012016

abstracts in scientific journals obtained from the semaval 2010 dataset and added with data from competent journals. Using the method of BERT, this research is expected to provide the best results of research using the NLP method, especially in the case of keyword searches.

2.Related Works

The approach describes related works for aspect extraction using deep learning algorithms for keywords and BERT methods based on study literature. In this case, the approach to describe this problem is using a deep learning algorithm. In deep learning, many methods can be developed in more depth. In its journey, the method can be trained to be able to classify the desired case. Deep learning classification is very dependent on big data to get optimal performance. Bidirectional Encoder Representations from Transformers (BERT) are capable and effective on a large number of datasets but will lack significant performance on a small number of trained datasets. In the study of M. R. Yanuar [3], adapting BERT to handle aspect extraction ABSA's aspect-based sentiment analysis assignment for Indonesian, especially in the tourist area domain. In M. Ramina et.al. [4], work implements a topic level with a level where the topic can be an idea, concept, or term that the user wants to know. Their research used NLP capabilities of the bidirectional encoder representation transformers (BERT) language model for generating a topic level.

In previous work of J. Yadav et.all [5] researched to calculate bullying comments for proposed solutions on social media based on the BERT model as a pre-trained BERT model with a single linear neural network layer on top as a classifier, which improves over the existing results. Furthermore, in Y. Iwasaki et.al [6] work developed an automatic abstractive text summarization algorithm.

Utilizing BERT to evaluating the model and also highlight future issues that could arise from the text generated as a result of training in the Japanese corpus. K. Ki et.al [7] proposed KoTAB (Korean Template-Based), a Korean template-based arithmetic solver with BERT, which is modified from TAB (Template-based Arithmetic Solver with BERT) framework that is used to solve English math word problems. KoTAB also solved Korean math word problems using two datasets, CC_Ko and IL_Ko, which were translated from English to Korean. He used four pre-trained BERT models that were used to fine-tune KoTAB; BERT-based, BERT-multilingual, Korean BERT WordPiece, and Korean BERT Morphology.

X. Zhao et.al [8] The combination model performance by combining BERT with several conversation models. The model investigates a better solution to using BERT: fixed weights or fine-tuning. Meanwhile, this paper describes new baseline methods based on BERT for the conversation generation task. We evaluate the proposed BERT-based models and baseline models on two popular dialog datasets: Ubuntu Dialog and Cornell Movie Dialog. The experiment results prove the effectiveness of incorporating BERT into conversation generation models.

3.Material and Methods

3.1. Dataset

In this study using the semaval 2010 dataset in its development the dataset is added by taking from other competent journal datasets, additional collections of other journals obtained from IEEE, in the dataset, there are 2 different file types, the first is a collection of journal abstracts. Scientific data and the second data is in the form of keywords from the abstract, the total scientific journal data used is 564 journals, which will be divided into 444 training (400 for train data, 44 for validation) and 10 testing data, for testing data obtained from IEEE journal.

3.2. BERT

BERT is a pre-trained language model which comprises a set of transformer encoders which represents the text at word and sentence level with the help of unsupervised training techniques like masked language modeling and next sentence prediction. Being a pre-trained model, BERT trained on 3300M words. BERT uses a transformer encoder with an attention mechanism to learn contextual relations between words. Transformer [9] in their native form consists of the encoder as well as decoder where the encoder learns the text input, and the decoder is tuned to conduct a specific task. As a language model, understanding the input text is the only important factor. Due to this reason, only transformer encoders are used in BERT. Rather than reading the text input sequentially like various directional models [9], the transformer encoder reads the entire sequence of words at once.

The BERT base model can be fine-tuned for text classification by simply adding a softmax classification layer on top, so it will predict the class of a given text sequence. It can be seen in Equal q.

$$p(c|H) = softmax(WH) \quad (1)$$

$$Z = softmax\left(Q\, x \frac{K^T}{\sqrt{d_k}}\right)V \quad (2)$$

Where Q is query matrix, K is a key matrix, V is a value matrix, d_k is the dimension of the key vektor, and Z is the attention score matrix of a single head. The attention score matrix is generated for every attention head, where score matrices are combined, and then reduce every input dimension of the feed-forward layer.

This functionality of the transformer helps generate context by calculating the relevance of each word concerning the presence of other words in the sentence. The level of contextual understanding is directly proportional to several transformer encoder layers. Fig. 1 shows BERT's architecture. BERT-Base, Uncased is used for the purpose which has 12 transformer layers, 768 hidden state, 12 attention heads, 110M parameters.

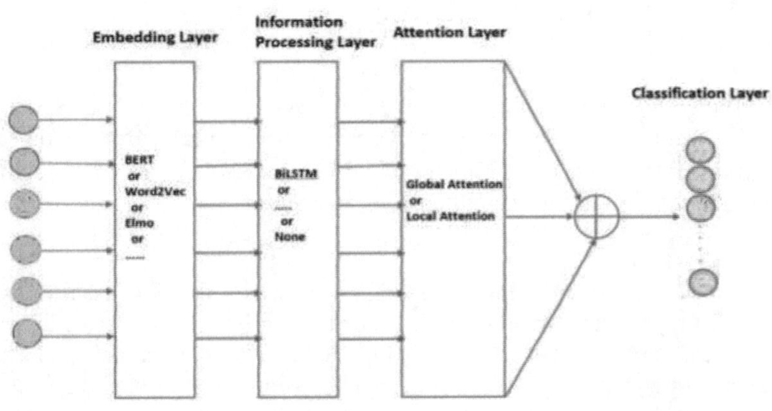

Figure 1. Attention-based model system design.

3.2.1. Fine Tunning BERT.

Understanding the input text is the only important factor in the BERT model because of the transformation of the encoder. Furthermore, the need in this work requires a transformer decoder. So, it will help this work to convert contextual embedding. The encoder is a pre-trained BERT language model, it takes less training time than a decoder.

IOP Conf. Series: Materials Science and Engineering 1175 (2021) 012016 doi:10.1088/1757-899X/1175/1/012016

Figure 2. BERT architecture for summarization.

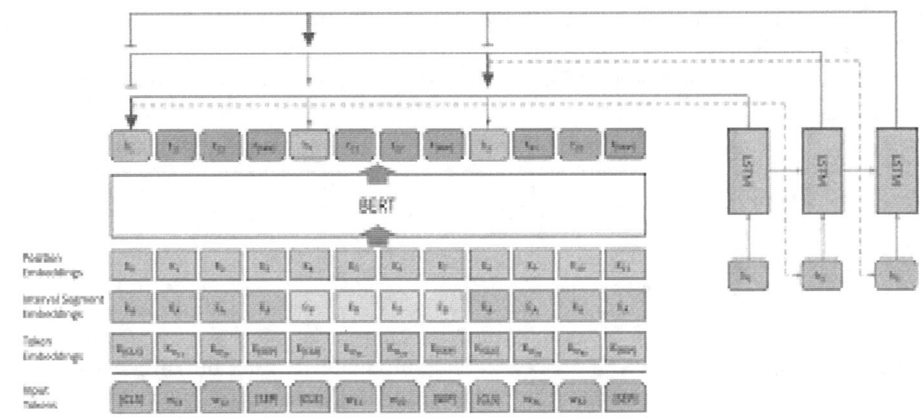

To make BERT adapt to a target task, it needs several factors, such as the preprocessing of long text since the maximum sequence length of BERT is 512, layer selection factor, and overfitting problem. A better optimizer with an appropriate learning rate will be desired. The lower layer of the BERT model may contain some information. So it can be fine-tuned with different learning rates. Following Howard and Ruder (2018) method, it split the parameters θ into $\{\theta^1,\ldots\ldots\theta^L\}$ where θ^l contains the parameters of the l-th layer of BERT. Then the parameters are updated as:

$$\theta_t^l = \theta_{t-1}^l - \eta^l.\nabla_{\theta^t} J(\theta) \quad (3)$$

Where η represents the learning rate of the l-th layer. And then set the base learning rate to η^L and use $\eta^{k-1}=\xi\cdot\eta^k$, where ξ is a decay factor and less than or equal to 1. But when $\xi<1$, the lower layer has a slower learning rate than the higher layer. If $\xi=1$ all layers have the same learning rate, so it will be equivalent to the regular stochastic gradient descent (SGD).

3.2.2. Further Pre-Tunning.
The BERT model is pre-trained for the general domain corpus. For the role of classifying text in a particular domain, such as movie reviews, the distribution of data may be different from BERT. Therefore, we can further pre-train BERT with masked language model and next sentence prediction tasks for domain-specific results. Three additional pre-training approaches are carried out :

1. Within-task pretraining, in which BERT is further pre-trained on the training data of a target task.
2. Pre-training in-domain, in which pre-training data is derived from the same domain of the aim mission. For example, there are many different sentiment classification tasks that have a similar distribution of data. We will further pre-train BERT on the cumulative training data for these tasks.
3. Cross-domain pretraining in which pre-training data is obtained from both the same and separate domains for the target task. We will discuss these diverse approaches to further pretraining.

3.2.3. Fine Tunning BERT.
Multitask Learning is also an important approach to knowledge exchange from a variety of various supervised tasks. Like [10], BERT is also used for text recognition in a multi-task learning system. Both functions are exchanged between the BERT layers and the embedding layer. The only layer that

IOP Conf. Series: Materials Science and Engineering 1175 (2021) 012016 doi:10.1088/1757-899X/1175/1/012016

does not shift is the final classification layer, which means that each work has a private classifier layer. Experimental analysis is being done in Sec.

4.Result and Discussion

In the experiment, classification testing was carried out using the BERT method and for this test using the 2010 semaval dataset with the entire dataset used in this study were 444 training (400 for train data, 44 for validation) and 10 test data for testing.

Performance in the classification is evaluated using performance metrics based on Train loss, validation accuracy, validation loss, and F1-score. The model with input data from the validation dataset. For validation accuracy, the calculated accuracy value of the validation dataset and prediction of the model with data input from the validation dataset and F1-score are used to calculate the average accuracy and recall. (4) (5) (6), TP, FP, FN and TN stands for true positive, false positive, false negative, and true negative

$$Recall = \frac{TP}{TP+FN} \quad (4)$$

$$Precision = \frac{TP}{TP+FP} \quad (5)$$

$$F1 - Score = 2x\frac{Precision\,x\,Recall}{Precision+Recall} \quad (6)$$

In building a strong word classifier, large amounts of data are needed to get good work results in the training process of this study using 500 epochs. The training results are shown in Table 1:

Table 1. Experimental Performance BERT with 500 EPOCH

epoch	***validation Accuracy***	***validati on loss***	***F1-score***	***Train loss***
500	0.96081	0.34035	0.58504	0.000868

To find out the toughness of the model, the training result model at 500 epochs was tested on 10 testing data. and calculated using equation (7). The test results show that the model obtains an average accuracy of 55.01%. The accuracy of the test results of each test data shown in Figures 5 and Figure 6 counted using this formula:

$$\text{Accuracy Test BERT} = \frac{\text{Success Rate}}{\text{count ground truth}}\; X\,100\% \quad (7)$$

The following is an example of the test results on one of the testing data "Achieving Reliable Sentiment Analysis in the Software Engineering Domain using BERT".

Table 2. Testing Result Each Testing Data

NO	Testing abstract	Accuracy
1	10.1109/ICSME46990.2020.00025	66.7 %
2	10.1109/ACCESS.2020.3030468	50.0 %
3	10.1109/TALE48000.2019.9225993	50.0 %
4	10.1109/ICSME46990.2020.00025	66.7%
5	10.1109/ICoICT49345.2020.9166407	50.0 %
6	10.1109/BIBM47256.2019.8983370	60.0%
7	10.1109/ICAIBD.2019.8837002	66.7%
8	10.1109/IUCC/DSCI/SmartCNS.00132	40.0%
9	10.1109/BigData47090.2019.9006511	50.0%
10	10.1109/ITNEC48623.2020.9085059	50.0%
	Avarage accuracy	55.01%

$$\text{Accuracy Test BERT} = \frac{2}{3} \text{x}100\% = 66.7\%$$

The success rate is obtained from the calculation of the success rate and divided by the total number of ground truth keywords and multiplied by 100%. Example of Bert's Test Results from one of the scientific journal test data with a test accuracy of 66.7%. The ground truth shown in Figure 4 and Figure 5 is the result of testing using models trained for 500 epochs.

Figure 4. Ground Truth

Input text
Researchers have shown that sentiment analysis of software artifacts can potentially improve various software engineering tools, including API and library recommendation systems, code suggestion tools, and tools for improving communication among software developers. However, sentiment analysis techniques applied to software artifacts still have not yet yielded very high accuracy. Recent adaptations of sentiment analysis tools to the software domain have reported some improvements, but the f-measures for the positive and negative sentences still remain in the 0.4-0.64 range, which deters their practical usefulness for software engineering tools.In this paper, we explore the potential effectiveness of customizing BERT, a language representation model, which has recently achieved very good results on various Natural Language Processing tasks on English texts, for the task of sentiment analysis of software artifacts. We describe our application of BERT to analyzing sentiments of sentences in Stack Overflow posts and compare the impact of a BERT sentiment classifier to state-of-the-art sentiment analysis techniques when used on a domain-specific data set created from Stack Overflow posts. We also investigate how the performance of sentiment analysis changes when using a much (3 times) larger data set than previous studies. Our results show that the BERT classifier achieves reliable performance for sentiment analysis of software engineering texts. BERT combined with the larger data set achieves an overall f-measure of 0.87, with the f-measures for the negative and positive sentences reaching 0.91 and 0.78 respectively, a significant improvement over the state-of-the-art.
Output text ***sentiment 0*** ***analysis 1*** ***bert 0***

Figure 5. Test Results

Input text
Researchers have shown that sentiment analysis of software artifacts can potentially improve various software engineering tools, including API and library recommendation systems, code suggestion tools, and tools for improving communication among software developers. However, sentiment analysis techniques applied to software artifacts still have not yet yielded very high accuracy. Recent adaptations of sentiment analysis tools to the software domain have reported some improvements, but the f-measures for the positive and negative sentences still remain in the 0.4-0.64 range, which deters their practical usefulness for software engineering tools.In this paper, we explore the potential effectiveness of customizing BERT, a language representation model, which has recently achieved very good results on various Natural Language Processing tasks on English texts, for the task of sentiment analysis of software artifacts. We describe our application of BERT to analyzing sentiments of sentences in Stack Overflow posts and compare the impact of a BERT sentiment classifier to state-of-the-art.

5. Conclusion

In this paper, an extensive experiment was conducted to investigate the approach to BERT, the task of text classification. There are several experimental findings: 1) The top layer of BERT is more useful for text classification; 2) With an appropriate layer-wise decreasing learning rate, BERT can overcome the catastrophic forgetting problem 3) In a training experiment, we can conclude that the greater the data and the smaller the training loss 4) From the results of the training trials on the epoch test 500, satisfactory test results are obtained in the training results by getting an validation Accuracy of 96.08%. 5) From the epoch 500 extract scientific journals and get a model from the training results. These results were tested with 10 abstract data. From the test results, calculations are made to get the results of the success rate.

6. References

[1] Tang M, Gandhi P, Kabir M A, Zou C, Blakey J and Luo X 2019 *Progress Notes Classification and Keyword Extraction using Attention-based Deep Learning Models with BERT. arXiv* [Online]. Available: http://arxiv.org/abs/1910.05786

[2] Devlin J, Chang M W, Lee K and Toutanova K 2019 BERT: Pretraining of deep bidirectional transformers for language understanding. *NAACL HLT 2019 - 2019 Conf. North Am. Chapter Assoc. Comput. Linguist. Hum. Lang. Technol. - Proc. Conf* **1** pp 4171 - 4186

[3] Yanuar M R and Shiramatsu S 2020 Aspect extraction for tourist spot review in Indonesian language using BERT. *Int. Conf. Artif. Intell. Inf. Commun. ICAIIC* pp 298 - 302 doi: 10.1109/ICAIIC48513.2020.9065263

[4] Ramina M, Darnay N, Ludbe C and Dhruv A 2020 Topic level summary generation using BERT induced abstractive summarization model. *Proc. Int. Conf. Intell. Comput. Control Syst. ICICCS 2020*. ICICCS pp 747 - 752 doi: 10.1109/ICICCS48265.2020.9120997

[5] Yadav J, Kumar D and Chauhan D 2020 Cyberbullying detection using pre-trained BERT model. *Proc. Int. Conf. Electron. Sustain. Commun. Syst. ICESC 2020*. ICESC pp 1096 - 1100 doi: 10.1109/ICESC48915.2020.9155700

[6] Iwasaki Y, Yamashita A, Konno Y and Matsubayashi K 2019 Japanese abstractive text summarization using BERT. *Proc. - 2019 Int. Conf. Technol. Appl. Artif. Intell. TAAI 2019* 1 doi: 10.1109/TAAI48200.2019.8959920

[7] Ki K, Lee D and Gweon G 2020 KoTAB: Korean template-based arithmetic solver with BERT. *Proc. - 2020 IEEE Int. Conf. Big Data Smart Comput. BigComp 2020* pp 279 - 282 doi: 10.1109/BigComp48618.2020.00-61

[8] Zhao X, Zhang Y, Guo W and Yuan X 2019 BERT for open-domain conversation modeling.

IEEE 5th Int. Conf. Comput. Commun. ICCC 2019 pp 1532 -1536 doi: 10.1109/ICCC47050.2019.9064414

[9] Sutskever I, Vinyals O and Le Q V 2014 Sequence to sequence learning with neural networks. *Adv. Neural Inf. Process. Syst* **4** January 2014 pp 3104 - 3112

[10] Liu X, He P, Chen W and Gao J 2020 Multi-task deep neural networks for natural language understanding. *ACL 2019 - 57th Annu. Meet. Assoc. Comput. Linguist. Proc. Conf* pp 4487 - 4496

IOP Conf. Series: Materials Science and Engineering 1175 (2021) 012017 doi:10.1088/1757-899X/1175/1/012017

Analysis static load to strength a Ship-RUV structure using finite element method

Budianto[1], I Sutrisno[2], M B Ramat[3], R Indarti[4], A W B Santosa[5]

[1,2,3,4] Politeknik Perkapalan Negeri Surabaya , Indonesia
[5]Universitas Diponegoro, Indonesia

E-mail: budianto.structure@gmail.com

Abstract. Conditions of accidents or marine transportation incidents can cause several conditions in the ship's part to experience structural deformation or a condition to experience cracks in the hull of the ship which causes the ship to experience flooding and sinking. The latest method of technology that is developing in the world today has used the unmanned dive method using RUV (Remotely Underwater Vehicle). It is a specific solution with Ship-RUV technology was developed by prioritizing simpler technology of size and function capacity but using reliable technology and materials. At first, the technology developed in Ship-RUV with design and analysis using finite element. Finite element used to solve the problem; it subdivides a large system into smaller, simpler parts with a numerical approach. Structural Ship-RUV is acceptable when actual stress 60.16 N/mm2 less than allowable stresses. In the initial ROV design structure analysis, it was found that the bending stress value reached 102.01 N / mm2 obtained from the calculation of cremona or mechanical in the stanchion structure. While in the analysis of new Ship-RUV model uses the Finite Element method and gets the value of the bending stress of 60.16N/mm2 is obtained which is equipped with a visual stress distribution

1. Introduction

Here Some of the ship's hull damage in operations include deformation, crack, fatigue, fouling, corrosion so that it requires an earlier process of investigating the hull so that there is no more damage to the hull. The number of sea transportation accidents, such as collision vessels, can cause some parts of the ship to be deformed or the condition of the ship's hull to rupture which. Some cases of ship crash in Indonesia were caused by collisions, that caused the ship's hull to break and sink. For example, the events in July 2018, the source of the Badan Nasional Pencarian dan Pertolongan (BASARNAS) information, where there were two ships colliding, which were types of tugboats (Tugboat) with the name of the TB Buana Express 10 ship which had 10 crew members serving the Kabaena (Southeast Sulawesi) route-Morowali Central Sulawesi) was hit by a passenger ship with the name of the KM Bunga Melati ship which had 17 crew members on the Tanjung Bakau (Southeast Sulawesi) route - Luwuk (Central Sulawesi) which crashed around the waters of Wawonii Island at coordinates 04 11 350 - 122 55 297E [8]. The investigation process in identifying leaking hulls must be carried out quickly and accurately before the ship sinks. Therefore, some special equipment is needed to deal with leaks. Some conventional tools for repairing ship leaks include axe and hammer, Rubber (it can be

Content from this work may be used under the terms of the Creative Commons Attribution 3.0 licence. Any further distribution of this work must maintain attribution to the author(s) and the title of the work, journal citation and DOI.
Published under licence by IOP Publishing Ltd

also a flip flops), water pumps and hoses, cement dried quickly, but it becomes inefficient when do not knowing the position of the leak on the hull of the ship. So that will cause a collapse [2].

Ship navigation is the most important part of the system on a ship. In the past, people always used celestial bodies such as constellations in determining the position of the direction of the voyage. This condition must require a deepening of knowledge that not everyone can understand in doing so. The ship navigation system can also function as a means of communication between other ships and can communicate with land. Such as Bio-fouling conditions in the hull of the ship, conditions where the thickness of the hull plate is reduced due to the presence of marine biota attached to the hull of the ship, these conditions can include deposit formation, encrustation, curding, deposition, scaling, scale formation, slagging, and bud formation [3]. Therefore, the Classification Bureau will always check the level of thickness of the ship's hull plate each year in terms of issuing the vessel's feasibility in operation at sea but docking must be carried out and this will require not a small cost. Therefore Ship-RUV technology is urgently needed to directly navigate hull conditions that are immersed in water, both due to fouling and even deformation conditions or cracks on the hull of the ship for a low-cost investigation process with a simpler arrangement of size and function but using technology and reliable material.

In the process of designing the Ship-RUV design methods are needed with the latest technology so that the optimum product results are obtained with a little row of residual waste material. The latest product design technology can control material requirements, structural strength, product optimization forms and can be more easily carried out by the process of designing the product. Some of the processes for designing Ship-RUV products with product design methods can be grouped into two parts, including Computer-Aided Design (CAD) is a computer device program for the process of drawing a product or part of a product component. Products that are designed can be represented by objects or symbols that have a specific purpose according to the standards that apply in the design. In CAD it can be 2D or 3D images. Computer-Aided Engineering (CAE) which is the use of Computer software in product design that assists technical analysis tasks. Including Finite Element Methods (FEM), Computer Fluid Dynamics (CFD) and optimization.

At first, the technology developed in Ship-RUV with design and analysis using finite element. The finite element method is used to solve problems, which divides large objects into smaller and simpler parts of elements by applying a numerical approach. Focus to strength a Ship-RUV Structure with analysis static load using FEM.The problem with the initial RUV structure is there are a lot of used the non-material standards in the stanchion structure, it will easily to occur the corrosion and crack because the structure is less strong caused the structural reinforcement is not well-defined for stress distribution.

2. Structure Analysis

Operational ships can experience damage to the ship's hull during operation, among others: deformation damage, fatigue, corrosion, etc. There are some hull conditions which cause it to be repaired as well as welding [7]. This is indeed the case: the vessel plate is hollow due to rust, the plate of the ship has holes with a large area, The thickness of the ship's plate has decreased to 20% from the standard thickness, The ship's plate is pushed into or deformed due to a collision with another object and Leather plate in the form of a double or double wave. So analysis structure is needed to ensure strength Ship-RUV structure.

2.1. Structure Design
In designing the ship wherein determining the main size and line plan includes several designs, among others: determine the main size of the ship, determine the design of the comparison ship, make Curve Section Area (CSA) and shape control body plan. Some designs of floating objects have a fairly high degree of difficulty. The amount of weight placed on the lightweight and death weight should not be more than the displacement of the planned vessel. In its development, the design of RUV can be done with three latest technologies, including: CAD technology and CAE technology. CAD is a computer device program for the process of drawing a product or part of a product component. In CAD it can be 2D or 3D images. CAE which is the use of Computer software in product design that assists technical analysis tasks. Including FEM, CFD, and optimization [4].

2.2. Remote Underwater Vehicle
For the development of underwater technology has less attention from the public. There are still many underwater activities carried out in conventional ways such as underwater observation, monitoring of dam fractures, surveys of underwater natural resources, search for victims of natural disasters or sinking ships and many more underwater activities so far carried out conventionally by humans [6]. Therefore, robots that can move freely in water are needed to help human tasks [5].

2.3. Fabrication
Fabrication of aluminum alloy 5083 while welding Aluminum 5083 to itself or another alloy from the same sub-group, the recommended filler metal is 5183. Other suitable fillers are 5356 and 5556.

Table 1 - Fabrication response for aluminum alloy 5083

Process Rating	Result
Workability – Cold	Average
Machinability	Poor
Weld ability – Gas	Average
Weld ability – Arc	Excellent
Weld ability – Resistance	Excellent
Bras ability	Poor
Bras ability	Poor

2.3.1. Temper Notation Code.
The most common tempers for aluminum 5083 are:

0 Annealed wrought alloy

H111 Some work hardening imparted by shaping processes but less than required for H11 temper.

H32 Work hardened and stabilized with a quarter hard tempers.

2.3.2. Applications of Aluminum Alloy 5083.
Aluminum 5083 is used in:

a. Shipbuilding
b. Rail cars
c. Vehicle bodies
d. Tip truck bodies

e. Mine skips and cages
f. Pressure vessels

2.4. Loads and Stress
In this case, the Ship-RUV has two kinds of loading, they are Static loads and gravity are a fixed load, both the magnitude (the intensity), the point of work and the direction of the line is fixed. In this case, it is the weight of the ship-RUV assuming the weight does not change.

The uniform load and the stresses that occur in the structure of Ship-RUV are analyzed by calculating the structure components. The calculation of the formula below to get the bending moment value can be used with the following equation:

$$Mmax = 1/12\ q\ \ell^2 \quad (1)$$

While the results of the calculation of bending stress are used to calculate the value of the maximum bending stress that occurs in a structure with existing loading, the following equation is shown:

$$\sigma = My / Ixx \quad (2)$$

2.5. Safety Factor
The safety factor in structural design can be used as an anticipation in the structure so that it can be precisely safe in its operation [1]. The development of planning methods based on security factors can be sorted as follows:
2.5.1. Ultimate Stress Method.
This method has two main methods: Plastic Method and Ultimate Stress Method.
2.5.2. Limit Stress Method.
Method of working stresses limit or often referred to as the stresses tension calculation, this is because of the metode used by stresses permits which is the melting stress (collapsed) material divided by a certain number called the safety factor (SF) like the following equation:

$$SF = \sigma u / \sigma i \quad (3)$$

SF: Safety Factor
σu: Melt / Ultimate Stresses
σi: Permission Stresses actual
Safety Factors can also be calculated from the BKI regulation for the construction of steel vessels as follows:
SF : Safety Factor
: 1.5 (in general)
: 1.2 (for construction that only accepts local load)
: 1.05 (for free static load combinations)

It can be concluded that the commonly used safety factor is 1.5. The safety factor used to analyzed the behavior of the ship-RUV is with a value of 1.5 for structure.

3. Methodology
Stresses analysis that occurs seen on the structure with used FEM, which can be calculated with the following steps design model. The design of the structure on the ship-RUV to be analyzed is modeled first with the AutoCAD software. Then it can export with ACIS file type (.sat). Then determination of the type of design by study. Based on BKI regulation Vol. II Year 2016 material used is marine used with maximum tensile pressure 400 N / mm2. In this step enter the mechanical properties to determine

the constraints possessed by the material. And Then meshing where model of navigation lampposts made in the form of three-dimensional given solid mesh form which is the recommended meshing type. Analysis of the behavior of the Ship-RUV is using fine mesh for accurate results The method allows for discontinuities, internal to the elements, in the approximation across the interface.Finally Analysis Structure, in the behavior analysis of Ship-RUV structure, post-processing analysis was performed with deformation and stresses simulations that occurred in the structure.

4. Result and Discussion

In the Ship-RUV structure analysis will be shown several stages and analysis process. So that the maximum results obtained in the analysis of Ship-RUV structure. This will be shown in several analysis designs by giving location of loading and constrain. Stress and strain will provide structural behavior when receiving operational loading.

4.1. Model 3-D Ship-RUV

In designing Ship-RUV modeling is done using Computer-Aided Design (CAD). With a modern design pattern where the design process begins with the making of 3D models then a modeling simulation using Computer-Aided Engineering is then carried out. Whereat the CAD stage it includes making a sketch area of a polyline and then forming a solid with an extrude pattern or revolved shape. For editing 3D models you can use a press pull or modify.

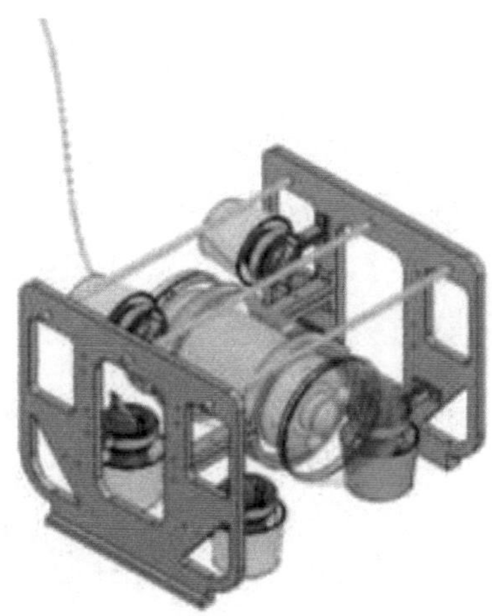

Figure 1. 3D Model of Ship-RUV

4.2. Simplify Menu for Structure Analysis

In this process, it is analyzed in the Ship-RUV structure and this is because the structural part is the component's constituent strength so that its strength is very calculated. It is unnecessary for simulation without affecting its production geometry (structure of ship RUV). In figure below shown simplify model of ship-RUV, It is not change native file if simplify menu to applied.

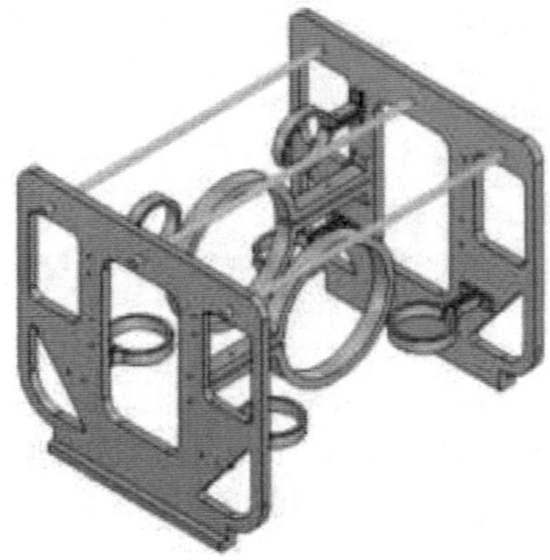

Figure 2. Simplify Model of Ship-RUV Structure

4.3. Material Ship-RUV

In the process of designing the Ship-RUV, use aluminum allow 5083 (marine used) material. Where the material is indeed designed for very corrosive ocean conditions and can reduce the marine biota attached to the Ship-RUV structure or in other words anti-fouling. The following is shown the chemical composition of the 5083 aluminum alloy material as follows:

Table 2. - Chemical composition for aluminum alloy 5083

Element	% Present
Si	0 . 4
Fe	0.4
Cu	0.1
Mn	0.4-1.0
Mg	4.0-4.9
Zn	0.25
Ti	0.15
Cr	0.05-0.25
Al	Balance

While it's the mechanical composition of the 5083-aluminum alloy is used as a reference for the material's ability to accept loads against the designed structure. In this case, it is taken at the elastic stretch limit, as the yield stresses limit where the material structure is in an inelastic condition which will be divided by the material standard factor and the structural safety condition factor. The following is shown the mechanical composition of the 5083-aluminum alloy material as follows:

Table 3. Mechanical properties for aluminum alloy 5083

BS EN 485-2:2008 Sheet 0.2 - 6.3mm Thick	
Property	Value
Proof Stress	125 Min MPa
Tensile Strength	275 - 350 MPa
Hardness Brinell	75 HB

**Properties above are material in the Soft O/H111 condition*

To show the physical properties of the 5083-aluminum alloy material, to determine the weight of the designed structure, after previously calculating the designed Ship-RUV volume. The following is shown the physical composition of the 5083-aluminum alloy material as follows:

Table 4. Physical c for aluminum alloy 5083

Property	Value
Density	2650 kg/m^3
Melting Point	570°C
Modulus of Elasticity	72 GPa
Electrical Resistivity	0.058x10^{-6} Ω.m
Thermal Conductivity	121 W/m.K
Thermal Expansion	25x10^{-6} /K

4.4. Constrain

Constrain is a placement condition in the operational conditions of Ship-RUV in maneuvering in sea-water. The constraint condition can be shown in the figure below, where the sagging maneuver hogging conditions are applied to the Ship-RUV structure to placement condition.

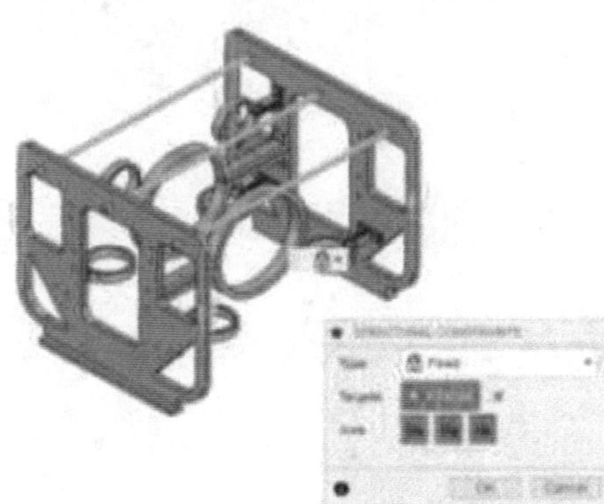

Figure 3. Constrain of Ship-RUV Structure

4.5. Constrain

Loading on Ship-RUV is emphasized on the trust that works on ship-RUV where it can be shown in the loading panel as follows:

a. Thruster after @ 100 N
b. Thruster bottom @ 100 N
c. Body @200 N

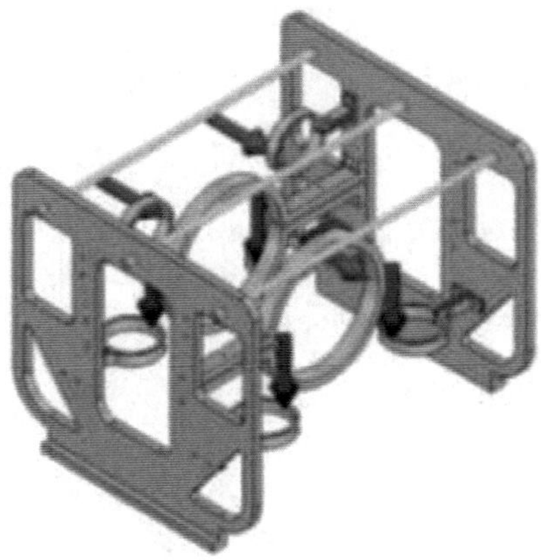

Figure 4. Load of Ship-RUV Structure

Contact (Connection Part) is Automatically calculates the contact types between all body and component that overlap or touch each other. To ensure all contacts are included in the simulation, run automatic contact before assigning manual contact. Note it can the later change manual types fro, the automatically generated one to the desired type. Also, you can suppress contacts to be excluded from the simulation. Meshing At this stage, the structure is divided into various constituent elements consisting of nodal and element parts which are engineering solver for the characteristic of the finite element method.

Figure 5. The Meshing of Ship-RUV Structure

Meshing properties result:

Table 5. Element Properties

Component	Value
164387 Tetrahedral	100.0% of elements
Face Angle min	1.98
Face Angle max	176
Dihedral Angle min	1.02
Dihedral Angle max	178
Worst shape ratio	92.2 on element 18385
Worst aspect ratio	9.23 on element 5246
Shortest edge	0.000162,
Longest	0.0334
Lowest collapse ratio	0.0157 on element 1908
Worst Jacobian ratio	3.89 on element 11317
Base mesh nodes	300083 nodes
Base mesh elements	164387 elements
Solver mesh nodes	300083 nodes
Solver mesh nodes	164387 elements
Job has finished successfully.	

4.6. VonMisses

Von Misses is a combined result of the value of the stress that occurs in the RUV ship structure where after the loading occurs.

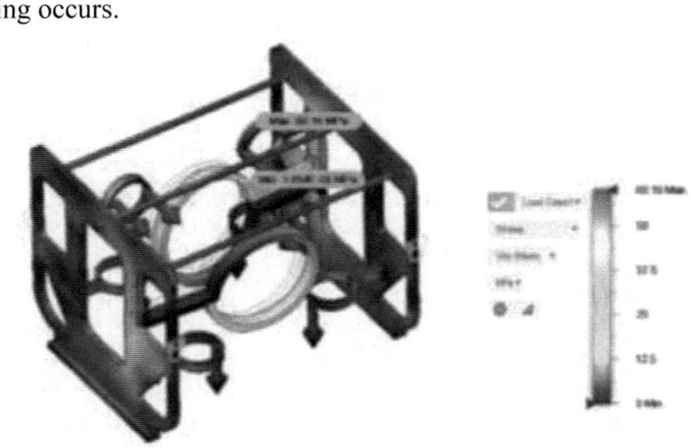

Figure 6. VonMisses Stresses of Ship-RUV Structure

IOP Conf. Series: Materials Science and Engineering 1175 (2021) 012017 doi:10.1088/1757-899X/1175/1/012017

Allowable stresses

$$\sigma_a = \frac{\sigma_{yield}}{SF.K} \tag{4}$$

Where :

σ_y = Yield Strength, N/mm^2

SF = Safety Factor

K = Factor Material

Result : structural is acceptable ($60.16 \leq \sigma a$)

σa = 125/1.5.1

= 83.33 N/mm^2

4.7. Deformation Result

Deformation is a structural behavior that occurs as a result of loading the structure that occurs. Where it's must be referred to the permissible limits of deformation that occurs.

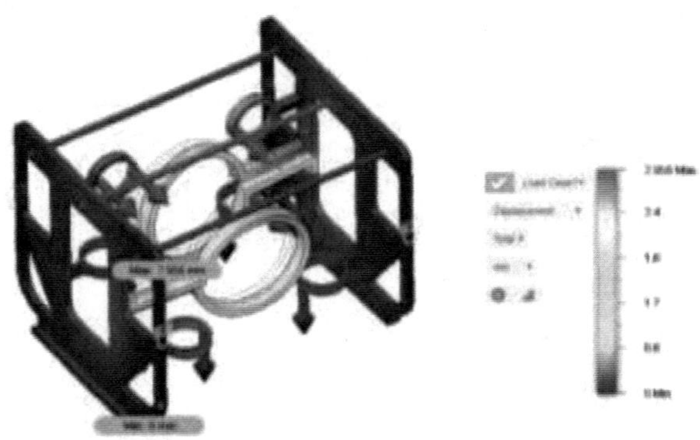

Figure 7. Deformation of Ship-RUV Structure

Allowable deformation

L/28 for both end continuous (5)

L/28 = 250/28

= 8.9 mm

Result of deformation structural is acceptable

5. Conclusion

Static Load is applied to strength a Ship-RUV Structure by Finite Element Analysis. It is a solution specific with Ship-RUV technology was developed by prioritizing simpler technology of size and strength of function capacity but using reliable structure technology and materials. At first, the technology developed in Ship-RUV with design and analysis using finite element. The finite element method is used to solve problems, which divides large objects into smaller and simpler parts of elements by applying a numerical approach. Focus to strength a Ship-RUV Structure with analysis static load using finite element method. Structural Ship-RUV is acceptable when actual stress 60.16 N/mm^2 less than allowable stresses 83.33 N/mm^2. The behavior of Ship-RUV structure can be analyzed by actual deformation. Maximum deformation is 2.9 mm less than allowable deformation with value 8.9 mm. In the initial ROV design structure analysis, it was found that the bending stress value reached 102.01 N / mm2 obtained from the calculation of cremona or mechanical in the stanchion structure.

while in the analysis of new Ship-RUV model uses the Finite Element method and gets the value of the bending stress of 60.16N/mm2 is obtained which is equipped with a visual stress distribution.

6. References

[1] Budianto M T, Wahyudi U, Dinata, Ruddianto and Eko M M P 2018 Strength analysis on ship ladder using finite element method. *Journal of Physics: Conference Series* **953** pp 1 - 8 012043
[2] *Marine Structures* 1991 U.S. Patent 5 018 471
[3] Mhatt G, LefranÃ E and Gilbert T 2012 *Finite Element Method* (John Wiley & Sons)
[4] M A H Koli 2015 Rancang bangun robot bawah air mini ROV (Remotely Operated Vehicles). *Jurnal Teknik Elektro Universitas Tanjungpura* pp 1-10
[5] Machine W 2009 *Autonomous Vehicles* 2nd Ed (Institute of Marine Technology: Bangladesh)
[6] Tutu S 2017 *Penyebab Kerusakan Kapal* Ed 1 (Surabaya)
[7] Wahyudi 2018 *Dua Kapal Tabrakan di Perairan Kendari: KM Bunga Melati 79 Tenggelam* (Kendari: Basarnas)
[8] Jami'in M A, Hu J, Marhaban M H, Sutrisno I and Mariun N B 2016 Quasi-ARX neural network based adaptive predictive control for nonlinear systems. *IEEJ Trans Elec Electron Eng* **11** pp 83-90
[9] Jami'in M A, Sutrisno I and Hu J 2012 Lyapunov learning algorithm for Quasi-ARX neural network to identification of nonlinear dynamical system. *IEEE International Conference on Systems, Man, and Cybernetics (SMC)* (Seoul, 2012) pp 3147-3152 doi: 10.1109/ICSMC.2012.6378275
[10] Jami'in M A, Sutrisno I and Hu J 2013 Deep searching for parameter estimation of the linear time invariant (LTI) system by using Quasi-ARX neural network. *The 2013 International Joint Conference on Neural Networks (IJCNN)* (Dallas, TX) pp 1-5 doi: 10.1109/IJCNN.2013.6707091
[11] Jami'in M A, Sutrisno I and Hu J 2015 Maximum power tracking control for a wind energy conversion system based on a quasi-ARX neural network model. *IEEJ Trans Elec Electron Eng* **10** pp 368-375
[12] Jami'in M A, Sutrisno I, Hu J, Mariun N B and Marhaban M H 2014 An adaptive predictive control based on a quasi-ARX neural network model. *13th International Conference on Control Automation Robotics & Vision (ICARCV)* (Singapore, 2014) pp 253-258 doi: 10.1109/ICARCV.2014.7064314
[13] Sutrisno I, Jami'in M A and Hu J 2014 Modified fuzzy adaptive controller applied to nonlinear systems modeled under quasi-ARX neural network. *Artif Life Robotics* **19** pp 22–26 https://doi.org/10.1007/s10015-013-0137-6
[14] Sutrisno I, Jami'in M A and Hu J 2014 An improved elman neural network controller based on quasi-ARX neural network for nonlinear systems. *IEEJ Trans Elec Electron Eng* **9** pp 494-501
[15] Sutrisno I, Che C and Hu J 2014 An improved adaptive switching control based on quasi-ARX neural network for nonlinear systems. *Artif Life Robotics* **19** pp 347–353
[16] Sutrisno I, Jami'in M A and Hu J 2012 Neural predictive controller of nonlinear systems based on quasi-ARX neural network. *18th International Conference on Automation and Computing (ICAC)* (Loughborough, 2012) pp 1-6
[17] Rinanto N and Khumaidi *Aplikasi Android-Rasspberry Pi Pada Kapal Tanpa Awak Untuk Pencarian Korban Kecelakaan Laut*
[18] Widodo R B et al Grasping and attached mode in human computer interaction in the study of mouse substitution. *Journal of Physics: Conference Series* **1196** 1 012034
[19] Sutrisno I, Jami'in M A, Hu J, Marhaban M H 2015 Self-organizing quasi-linear ARX RBFN modeling for identification and control of nonlinear systems. *54th Annual Conference of the Society of Instrument and Control Engineers of Japan (SICE)*. IEEE

[20] Jami'in M A, Sutrisno I and Hu J 2014 Nonlinear adaptive control for wind energy conversion systems based on quasi-ARX neural networksmodel. *Procedings of the International Multi Conference of Engineers and Computer Scientists.* March 12-14, 2014 (Hong Kong)
[21] Sutrisno I 2009 *Pemrograman Komputer dengan Software Matlab* (Surabaya: ITS Press)

Acknowledgement

Thanks to Ministry of Research, Technology and Higher Education (Kemenristekdikti) on the grantees of research schemes applied by number contract: 1616.11/PL19/LT/2019 and several lectures in Politeknik Perkapalan Negeri Surabaya (PPNS) with sharing knowledge until finish this research.

IOP Conf. Series: Materials Science and Engineering 1175 (2021) 012018 doi:10.1088/1757-899X/1175/1/012018

A review of concrete durability in marine environment

W D Pratiwi[1], F D D Putra[2], Triwulan[3], Y Tajunnisa[4], N A Husin[5] and K D Wulandari[6]

[1,2,6] Shipbuilding Institute of Polytechnic, Surabaya, Indonesia
[3] Civil Engineering, Institut Teknologi Sepuluh Nopember, Surabaya, Indonesia
[4,5] Civil Infrastructure Engineering, Institut Teknologi Sepuluh Nopember, Surabaya, Indonesia

E-mail: wiwik.pratiwi@ppns.ac.id

Abstract. There are many concrete constructions in marine environments. Due to its salinity content, seawater is aggressive, both towards concrete and reinforcement. Concrete interacts with seawater which causes deterioration. The chemical attack occurs at least through 3 common mechanisms: sulfate-attack, carbonation, and chloride-induced corrosion of reinforcement. The chemical resistance of concrete is determined by the porosity and chemical composition of the concrete. Partly subsitution of cement with supplementary cementitious materials (SCM) known as an effective way to create the less permeable concrete. Concrete with fly ash as SCM has excellent potential to be applied to marine construction in Indonesia because the abundant availability of fly ash in Indonesia and the presence of aluminum species in fly ash provide an opportunity to increase binding capacity to chloride.

1. Introduction

Concrete is a material that is widely used in marine construction, for examples are docks, piers, breakwaters, LNG terminals, offshore windmills, and many other structures. The durability of a structural concrete in a specified marine environment likely to depend upon the structural factor (geometries, crack) and material factors (mixture components and its composition, curing, etc). Degradation of reinforced concrete can be classified into: physical (freeze-thaw cycles, fire); mechanical (erosion, abrasion); chemical; biological (biogen attack); structural (overloading). The main chemical damage mechanisms of reinforced concrete are caused by sulfate attack, carbonation, calcium leaching and chloride-induced corrosion.

The hardened cement paste always has capillary pores which decrease in volume as the water to cement ratio (w/c) drops and the degree of hydration increases. The same is true of concrete, but in concrete tends to be more porous as the presence of the aggregates. The existence of these capillary pores creates permeability in concrete. At high porosity, the interconnected pore system connects the concrete surface to the bulk of concrete. Permeability increases sharply if the porosity exceeds 25%, which is obtained from the w / c 0.45 and 75% degree of hydration [1].

Concrete permeability causes ions in seawater to enter the structure or vice versa leaching the constituent components of the concrete to sea water. Four basic mechanisms of the passage of fluid

Content from this work may be used under the terms of the Creative Commons Attribution 3.0 licence. Any further distribution of this work must maintain attribution to the author(s) and the title of the work, journal citation and DOI.
Published under licence by IOP Publishing Ltd

and ions through concrete are: capillary suction, due to capillary action inside capillaries of cement paste, permeation, due to pressure gradients, diffusion, due to concentration gradients, and migration, due to electrical potential gradients [1]. The transport rate of ions or fluid depend on mechanism, concrete porosity, and the binding of transported substances by the hydrated cement paste as well as on the environmental conditions.

Many studies have been carried out to improve the impermeable properties of concrete, for example by adding clay brick powder [2], regulating water to cement ratio [3], fly ash substitution [4–6] and blast furnace slag subtisution [7].

Concrete exposed to seawater is deteriorated due to chemical and physical interactions. The chemical failures are majorly caused by sulfate attack, carbonation and chloride-induced corrosion. While mechanical action and the growth of marine biota affect concrete degradation, however, this paper is focused on the durability against chemical attack. The main objective of this paper is to study the chemical breakdown mechanism of concrete and the critical properties of concrete that affect its resistance. Further research needs to be identified to produce concrete that is resistant to seawater attacks by taking into account the availability of materials in Indonesia.

2. Sulfate Attack

The aggressiveness of seawater to concrete is influenced by several factors, including the composition of seawater, climate, waves, and marine zones (immersion, tidal, splash or spray zones). The main ions present in seawater are Na^+, Mg^{2+}, Cl^- and SO_4^{2-}, together with the small quantities of K^+, Ca^{2+}, HCO_3^- and Br^- [8]. The salinity of seawater varies widely, but it can be assumed that an average of 3.5% (35 g / L dissolved salt) [9]. The permeability of the concrete and the compositions of the hydrated phases of the cement are the main factors of concrete materials governing the sulfate reaction. The hydration products of the tricalcium aluminate (C_3A) are known as the susceptible phase to sulfate attack [10].

Sulfate ions from the environment attack the concrete and are known as external sulphate attack (ESA). Sulfate attacks originating from concrete components known as internal sulfate attacks (ISA) or delayed ettringite formation (DEF) are not included in this paper. External sulfate attack includes all chemical reactions from external sulfate sources that cause concrete damage regardless of the type of cation. The sulfate attack mechanisms consists of [8,10]:

2.1. The reaction of sulfates with the constituent components of the cement paste produces gypsum or is known as gypsum attack. The two common reactions are:

2.1.1. The reaction of sodium sulfate with calcium hydroxide and water result gypsum and sodium hydroxide.

$$Na_2SO_4+Ca(OH)_2 + 2H_2O \rightarrow CaSO_4.H_2O + 2NaOH \quad (1)$$

2.1.2. The reaction of $MgSO_4$ with $Ca(OH)_2$ produce gypsum and magnesium hydroxide.

$$MgSO_4 + Ca(OH)_2 + 2H_2O \rightarrow CaSO_4 \cdot 2H_2O + Mg(OH)_2 \quad (2)$$

The gypsum produced by those reactions is believed to damage concrete by increases in overall solid volume. In addition, the reduction of $Ca(OH)_2$ lead to depasssivation of reinforcing steel as decreasing the alkalinity of pore solution. Magnesium hydroxide (eq. 2) has very low solubility so it tends to form deposits on the concrete surface. These deposits can protect the next attack if the construction is submerged [8,10].

IOP Conf. Series: Materials Science and Engineering 1175 (2021) 012018 doi:10.1088/1757-899X/1175/1/012018

2.2. *Expansion occurs due to the formation of ettringite (called a sulfoaluminate attack)*, include two reactions:

2.2.1. *Gypsum reacts with calcium aluminate hydrate plus water to form ettringite and calcium hydroxide.*

$$3(CaSO_4.2H_2O) + 4CaO.Al_2O_3.19H_2O + 8H_2O \rightarrow 3CaO.Al_2O_3.3CaSO_4.32H_2O + Ca(OH)_2 \quad (3)$$

2.2.2. *The reaction of sodium sulfate with calcium monosulfoaluminate, calcium hydroxide and water produce ettringite and sodium hydroxide.*

$$2Na_2SO_4 + 3CaO.Al_2O_3.CaSO_4.12H_2O + 2Ca(OH)_2 + 2H_2O \rightarrow 3CaO.Al_2O_3.3CaSO_4.32H_2O + 4NaOH \quad (4)$$

The formation of ettringite (eq. 3, 4) causes volume expansion, hence generate the cracks [11–13]. However, interactions with chloride prevent ettringite from forming [8].

2.3. *Decomposition of calcium-silicate-hydrate, calcium aluminatehydrate, calcium monosulfoaluminate, resulting in the loss of cementing properties (paste cohesion, paste-to-aggregate bond, strength, and durability*

The reaction of $MgSO_4$ with calcium aluminate hydrate

2.3.1. *Forms gypsum, magnesium hydroxide and aluminum-hydroxide.*

$$3MgSO_4 + 3CaO.Al_2O.36H_2O + 6H_2O \rightarrow 3(CaSO_4.2H_2O) + 3Mg(OH)_2 + 2Al(OH)_3 \quad (5)$$

2.3.2. *The reaction between magnesium sulfate and calcium monosulfoaluminate hydrate produce gypsum, magnesium hydroxide, and aluminum hydroxide.*

$$3MgSO_4 + 3CaO.Al_2O_3.CaSO_4.12H_2O + 2H_2O \rightarrow 4(CaSO_4.2H_2O) + 3Mg(OH)_2 + Al_2O_3.3H_2O \quad (6)$$

2.3.3. *The reaction of magnesium sulfate and calcium silicate hydrate result gypsum, magnesium hydroxide and silica gel.*

$$3MgSO_4 + 3CaO.2SiO_2.nH_2O + zH_2O \rightarrow 3(CaSO_4.2H_2O) + 3Mg(OH)_2 + 2SiO_2.xH_2O \quad (7)$$

2.3.4. *Magnesium sulfate reacts with calcium silicate hydrate and calcium hydroxide to form magnesium silicate and gypsum.*

$$4MgSO_4 + CaO.SiO_2.H_2O + (4-x)\,Ca(OH)_2.nH_2O \rightarrow 4MgO.SiO_2.8H_2O + 4CaSO_4.2H_2O \quad (8)$$

2.4. *Thaumasite sulfate attack, namely sulfate attacks together with carbonate to produce thaumasite*

Thaumasite is a pulpy soft mass and has no binding ability. This attack often occurs at low temperatures (<10°C)

$$3Ca^{2+} + SiO_2^{3-} + CO_3^{2-} + SO_4^{2-} + 15H_2O \rightarrow$$
$$3CaO \cdot SiO_2.CO_2 \cdot SO_3 \cdot 15H_2O \quad (9)$$

2.5. *Salt recrystallization*

The ions present in the environment have penetrated the concrete and then react with various possibilities as mentioned above. When the ions, generally sodium and sulfate, reach the surface then recrystallization occurs. This recrystallization often happens when the concrete experiences changes in temperature and humidity, for example in areas that subjected to drying-wetting cycles or tidal zones. Some authors classify it as a physical attack, however, disallowed by the other since the process involves chemical reactions [8].

3. Chloride-Induced Corrosion of Concrete Reinforcement

Chloride-induced corrosion is assumed as the foremost origin of reinforced concrete damage. Many studies have been carried out to determine the values for critical chloride content (C_{crit}) in reinforced concrete, i.e. the chloride concentration value on the surface of reinforcing steel which causes corrosion initiation. According to European standards, EN 206-1, C_{crit} value is 0.2 or 0.4% by weight of cement, depending on the type of cement [14]. In contrast, some experiment state the higher value, for example Kwon [15] reported C_{crit} of 2.5%. The C_{crit} value is the most decisive input parameter in the service life probability modelling.

Chloride from the environment, which is absorbed on the concrete surface enters the deeper zone by diffusion. The diffusion coefficient is the main parameter to describe the diffusion process that occurs.

The mechanism of chloride attack on reinforcing steel occurs in the following steps [1].

(i) Chloride ion penetrates the oxide film on the steel surface at the defect site. The penetration of chloride ion is stronger than other ions, therefore termed as the aggressive ion.
(ii) Chloride ion is adsorbed on the metal surface. At the same time, dissolved oxygen and hydroxyl ion are adsorbed on the metal surface;
(iii) The ferrous-cation formed from corrosion reacts with chloride ions to form a soluble complex of iron chloride. This compound leaving the metal surface attacks the protective layer of $Fe(OH)_2$ so that corrosion continues.

The C_{crit} value is influenced by the pH of the concrete or concentration of hydroxyl ions in the pore solution, steel potential, and the presence of voids on the steel/concrete interface. The concentration of hydroxyl ions depends on the type of cement mixture. The risk of corrosion attack increases when the $[Cl^-]/[OH^-]$ ratio increases to a certain value. However, several authors have shown different critical values [1].

The substitution of cement with SCM has two-fold effects. The first has the effect of reducing porosity, and the second increases the binding capacity of chloride. The binding capacity behaviour for cement paste with metakaolin, silica fume, fly ash, and slag substitution has been studied. The highest binding capacity is found in pastes with metakaolin. This fact can be understood since the formation of Friedel's salt ($FC_3A.CaCl_2.10H_2O$) requires aluminium, while metakaolin is a mineral with a high aluminium content. The use of fly ash with different calcium content did not show a significant

IOP Conf. Series: Materials Science and Engineering 1175 (2021) 012018 doi:10.1088/1757-899X/1175/1/012018

difference. However, apart from the type of mineral, the w/c factor also plays an important role. The increase in binding capacity due to mineral substitution was detected at w/c 0.5, while at w/c 0.3, the binding capacity change was only slightly. It is suspected that an increase in w/c provides the availability of free Al for the formation of Friedel's salt [16].

In another study, concrete with 56 and 58% fly ash was exposed to the marine environment for 24 years. It was found that the cement substitution with fly ash resulted in resistance to chloride ion penetration. However, to get the best results, it is necessary to take synergic steps of SCM substitution, low w/c ratio (not more than 0.4) and sufficient cover depth. [6].

The combination of chloride and magnesium sulfate attack was studied on the three kinds of mortar made of Portland cement, sulfate resisting cement and a blended of 50% OPC -50% slag. Substitution with slag prevents the formation of brucite ($Mg(OH)_2$) hence there are no protective properties on the surface [17].

4. Carbonation

Carbonation take places when carbon dioxide in the environment infiltrates into the concrete to react with calcium hydroxide giving calcium carbonate:

$$CO_2 + Ca(OH)_2 + \rightarrow CaCO_3 + H_2O \qquad (10)$$

Reduction of $Ca(OH)_2$ causes an increase in porosity and causes a decrease in pH. Under normal conditions, the pH of the concrete pore solution is in the range 13-14 so that the reinforcing steel is in a passive state. In the case of carbonization, the pH can drop close to neutral.

Water to cement ratio (w/c) is the main factor controlling the resistance against carbonation of normal concrete. When chloride ions ingress and carbonation occur at the same time in concrete, chloride ions become soluble due to the deterioration of cement hydrates caused by carbonation. In addition, the chloride bound in the form of calcium chloroaluminate hydrate is released into the pore solution so that it becomes more aggressive [1].

Carbonation is affected by moisture. In both completely dry and wet environments, carbonation does not occur. In marine structures, the splash zone and above water are at risk of carbonation attacks. The condition for the attack is getting worse if there is chloride from both concrete components and the environment [1].

Research to determine the resistance of concrete to gas and water permeability was carried out on high-volume fly ash (HVFA). A 50% fly ash mixture by mass and a water-to-binder ratio of 0.4 is obtained to have water and gas permeability lower than normal concrete, provided that fly ash has a fairly low LOI (loss on ignition), 3.5% and high refinement (19% retained) on a 45 μm sieve [18].

The leaching resistance of concrete is sensitive to w/c and the paste composition. The concrete undergoes leaching due to the presence of portlandite in the paste. Furthermore, the leached concrete is susceptible to sulfate attack. High C3S cement substitution with 30% FA had an increased impact on leaching resistance and sulfate attack. Substitution of cement with 62% GGBS (ground granulated blast-furnace slag) gives the best results [12]. However, different results were found by Ashish [19], the slag substitution of 20% affected increasing durability, but an increase in substitution of 40% or more resulted in a decrease in durability.

5. Discussion

Marine concrete must be resistant to sulfate attack, carbonation, calcium leaching and chloride penetration. Physically, low permeability properties are the main requirement for concrete to have

IOP Conf. Series: Materials Science and Engineering 1175 (2021) 012018 doi:10.1088/1757-899X/1175/1/012018

resistance to chemical attack. However, the chemical composition of concrete is also a determinant besides its permeability.

Permeability is controlled by the w/c ratio and the volume of hydrated cement. The higher the w/c, the larger the pore volume and the increased permeability. Carbonization and calcium leaching are attacks against $Ca(OH)_2$ from hydration. Carbonized and calcium leaching concrete has a higher porosity, making it more sensitive to attack by other chemical species, such as sulfate or chloride. Moreover, calcium leaching causes a higher supply of calcium ions in the pore solution thereby increasing the rate of reaction in chemical attack [20].

Concrete with fly ash or GGBS substitution has better permeability than normal concrete. Mittermayr [21] found that concrete made with CEM IIIB cement (Blast furnace cement), fly ash and calcium carbonate with low w/c have high resistance to sulfate attack. The use of calcium carbonate, together with fly ash provides a synergistic effect in increasing cement hydration. Calcium carbonate chemically reacts with the aluminate in fly ash to form a carboaluminate phase, whereas physically, the calcium carbonate surface provides an additional site for the C-S-H phase deposition. The chemical similarity of calcium carbonate and C-S-H enhance the growth of C-S-H calcium carbonate particles [22,23].

Efforts to reduce pores are also carried out by adding inhibitors based on natural ingredients. Bambusa Arundinacea is reported to have the ability to replace nitrite and amine-based corrosion inhibiting admixtures for durable concrete structures due to its pore-blocking effects [24]. Meanwhile, [25] used silica fume to reduce pores and increase chemical resistance.

The resistance of concrete to chloride attack requires special attention. Sulfate-resistant cement is usually designed to have a low alumina content. This is because the alumina reacts with the sulfate to form an expansive ettringite which increases the risk of cracking. On the other hand, alumina is a species necessary to bind chloride to form Friedel's salt. The binding capacity of a concrete determines its resistance to chloride. And almost the same as the resistance to sulfate, concrete with SCM has better properties than normal concrete. If in general, the w/c is designed to a minimum to produce concrete with small pores, this is not the case with chloride resistant. The value of w/c needed is not too small, generally around 0.4 to ensure the availability of alumina which is ready to bind with chloride ions.

The porosity of the concrete is determined by w/c and the hydrated phases. Several studies have recommended blended cement with GBFS to produce a mixture that is durable in seawater. However, there is a controversy that says the substitution of GBFS is not suitable for environmental conditions containing magnesium sulfate and chloride.

Concrete with cement substitution using fly ash is considered suitable for applications in the marine environment in Indonesia. This is due to the abundant availability of fly ash in Indonesia. In 2020, fly ash and bottom ash from the steam-power plants in Indonesia be around 13.5 million tons, and only 10% have been used. Another reason is that the substitution of fly ash, with any calcium content, is thought to be effective in producing concrete that is able to withstand the attack of seawater, especially chloride. This is because fly ash has an aluminium content for the formation of Friedel's salt so that the chlorine is bound to the concrete.

The drawback of fly ash is its relatively low reactivity so that the substitution of fly ash at a relatively high level has an impact on decreasing compressive strength. Selection of fly ash with high fineness and high morphology [29] provides an opportunity to produce good properties as SCM, as well as adding activator to increase its reactivity [26–28].

6. Conclusion

Marine concrete has a risk of chemical degradation against sulfates, carbonates and chlorides attacks. For reinforced concrete, chloride ion attack is critical since it generates corrosion of the reinforcement.The chemical resistance of concrete is determined by the porosity and the chemical composition of the concrete. The aluminium content is required for concrete to increase the binding capacity against chloride. Concrete with fly ash as SCM has excellent potential to be applied to marine construction in Indonesia because the abundant availability of fly ash in Indonesia and the presence of aluminium species in fly ash provide an opportunity to increase the binding capacity to chloride.

7. References

[1] Bertolini L, Elsener B, Pedeferri P and Polder R 2004 *Corrosion of Steel in Concrete Prevention, Diagnosis, Repair* (Weinheim: WILEY-VCH Verlag GmbH & Co. KGaA)

[2] Castillo M, Hernández K, Rodriguez J and Eyzaguirre C 2020 Low permeability concrete for buildings located in marine atmosphere zone using clay brick powder. *IOP Conf. Ser. Mater. Sci. Eng.* **758**

[3] Kuhail Z and Shihada S 2001 *Permeability and Chloride Penetration of Concrete Subjected to Islam. Univ. gaza*

[4] Dinakar P, Babu K G and Santhanam M 2008 Durability properties of high volume fly ash self compacting concretes. *Cem. Concr. Compos.* **30** 880–6

[5] Nath P and Sarker P 2011 Effect of fly ash on the durability properties of high strength concrete. *Procedia Eng.* **14** 1149–56

[6] Moffatt E G, Thomas M D A and Fahim A 2017 Performance of high-volume fly ash concrete in marine environment. *Cem. Concr. Res.* **102** pp 127–35

[7] Choi S Y, Choi Y S and Yang E I 2016 Characteristics of calcium leaching resistance for concrete mixed with mineral admixture. *J. Korea Inst. Struct. Maint. Insp.* **20** pp 59–67

[8] Skalny J, Marchand L and Odler I 2002 *Sulfate Attack on Concrete* (London and New York: Spon Press)

[9] Thomas M 2016 The durability of concrete for marine construction. *Marine Concrete Structures* (Elsevier Ltd) pp 151–70

[10] Thomas M D A and Skalny J 2006 Chemical resistance of concrete. *Significance of Tests and Properties of Concrete and Concrete Making Materials. STP 169D* ed J F Lamond and J H Pielert (West Conshohocken, PA: ASTM International) pp 253–73

[11] Robbins D B and Lebel L B 2016 Ports 2016: Port Planning and Development *Developing State-of-the-Art Marine Concrete Repair Daniel* pp 441–50

[12] Rozière E, Loukili A, El Hachem R and Grondin F 2009 Durability of concrete exposed to leaching and external sulphate attacks. *Cem. Concr. Res.* **39** 1188–98

[13] Merida A and Kharchi F 2015 Pozzolan concrete durability on sulphate attack. *Procedia Eng.* **114** 832–7

[14] CEN 2013 BS EN 206: 2013 Specification , performance , production and conformity *Br. Stand.*

[15] Kwon S J 2017 Current trends of durability design and government support in South Korea: Chloride attack. *Sustain.* **9**

[16] Thomas M D A, Hooton R D, Scott A and Zibara H 2012 The effect of supplementary cementitious materials on chloride binding in hardened cement paste. *Cem. Concr. Res.* **42** pp 1–7

[17] Maes M and De Belie N 2017 Influence of chlorides on magnesium sulphate attack for mortars with Portland cement and slag based binders. *Constr. Build. Mater.* **155** 630–42

[18] Van Den Heede P, Gruyaert E and De Belie N 2010 Transport properties of high-volume fly ash concrete: Capillary water sorption, water sorption under vacuum and gas permeability.

Cem. Concr. Compos. **32** 749–56
[19] Ashish D K, Singh B and Verma S K 2016 The effect of attack of chloride and sulphate on ground granulated blast furnace slag concrete. *Adv. Concr. Constr.* **4** 107–21
[20] Qin S, Zou D, Liu T and Jivkov A 2020 A chemo-transport-damage model for concrete under external sulfate attack. *Cem. Concr. Res.* **132** 106048
[21] Mittermayr F, Rezvani M, Baldermann A, Hainer S, Breitenbücher P, Juhart J, Graubner C and Proske T 2015 Sulfate resistance of cement-reduced eco-friendly concretes. *Cem. Concr. Compos.* **55** 364–73
[22] Pratiwi W D, Triwulan, Ekaputri J J and Fansuri H 2020 Combination of precipitated-calcium carbonate substitution and dilute-alkali fly ash treatment in a very high-volume fly ash cement paste. *Constr. Build. Mater.* **234** 117273
[23] Weerdt K De, Kjellsen K O, Sellevold E and Justnes H 2011 Synergy between fly ash and limestone powder in ternary cements. *Cem. Concr. Compos.* **33** 30–8
[24] Abdulrahman A S, Ismail M and Hussain M S 2011 Inhibiting sulphate attack on concrete by hydrophobic green plant extract. *Adv. Mater. Res.* **250–253** 3837–43
[25] Alam B, Afzal S, Akbar J, Ashraf M, Shahzada K and Shahab M E 2013 Mitigating sulphate attack in high performance concrete. *Int. J. Adv. Struct. Geotech. Eng.* **02** 11–5
[26] Donatello S, Fernández-Jimenez A and Palomo A 2013 Very high volume fly ash cements. Early age hydration study using Na2SO4 as an activator *J. Am. Ceram. Soc.* **96** 900–6
[27] Pratiwi W D, Fansuri H, Ekaputri J J and Triwulan 2017 Effect of activators on strength of hybrid alkaline cement. *IOP Conference Series: Materials Science and Engineering*
[28] Pratiwi W D, Triwulan, Ekaputri J J and Fansuri H 2018 Investigation of one-part hybrid alkaline cement with heat-treated aluminosilicate Activator Better, Smarter, Stronger. *Proceeding for the 2018 fib Congress Held in Melbourne Australia, October 7-11, 2018* (Melbourne: Fédération internationale du béton (fib)
[29] Pratiwi W D 2019 Hubungan Morfologi , Ukuran partikel dan keamorfan Fly Ash dengan kuat tekan pasta. *Seminar Nasional MASTER 2019. Maritim, Sains dan Teknologi Terapan* (Politeknik Perkepalan Negeri Surabaya: Politeknik Perkapalan Negeri Surabaya) pp 92–8

IOP Conf. Series: Materials Science and Engineering 1175 (2021) 012019 doi:10.1088/1757-899X/1175/1/012019

Design of twist tool by Ulrich Method

M Hakam[1], S B Astsalisa[2], T A Setiawan[3]
[1,2,3]Teknik Desain dan Manufaktur, Politeknik Perkapalan Negeri Surabaya

E-mail: m_hakam@ppns.ac.id

Abstract. This research was conducted in an Outdoor Furniture which is one company that manufactures various types of outdoor furniture. Various types of chairs, sofas and tables are produced here through various types of manufacturing processes. In a sofa product, a component requires a manufacturing process that is quite complex and new in this company, namely the process of twist hollow pipes. This component is made from aluminum which measures 120mmx30mmx708mm. Production requests for products that require these components are counted in large quantities. However, the absence of tools that can be used to process the pipe is an obstacle for workers to produce these components. At present, the working process is carried out by adjusting or beating pipes in accordance with component design requirements. To overcome this, the design of a tool to twist hollow pipes was made specifically for this component. This tool was made using the Ulrich method for the problem analysis process, making 3 alternative concepts until there was a chosen concept, namely third concept. Then the tool isproduced and in accordance with the needs of the company and is able to facilitate the working of product components with a 95% success rate of twisting.

1. Introduction

The manufacturing industry is one of the fields that is being intensively developed in this era. The manufacturing sector in Indonesia is able to contribute up to 20% of Gross Domestic Product (GDP). The furniture industry is one of the industries that experienced the highest production growth in 2019 against 2018 with a percentage increase of 6.63%. This research was conducted at an Outdoor Furniture Company which is a company engaged in the manufacturing of outdoor furniture industry. Today, this company receives an order to work according to the design of one of the buyers which is somewhat different from the previous products. This product has one component that requires twisting / twisting, or making angles at several edges. Although the needs of producing this product are quite a lot, the company does not yet have the tools that make it easy for the twisting process. Likewise, the twist tools on the market are only able to work on components measuring less than 8mm x 8mm, while the company's needs require a tool to twist 120mm x 30mm components. From these problems, the authors plan to make a twist tool for the manufacturing process of these hollow pipe components. A twist tool specifically for ordering components of this product. This tool is considered very necessary because of the high production requirements for this, as well as the complexity of the work if the manufacturing process is done without the help of tools. For the method to be used in this study is the Ulrich method. The design of this twist tool is expected to make it easier to work with a large number of components in the company.

Some considerations made in this study were carried out with several types of analysis such as identification of existing products, components, torque strength, anthropometry, and direct analysis of related software. Twist or twisting tool is a tool used to twist the profile according to the angle of rotation determined at the time of twisting. Nowadays twisting tools are often used to make iron spiral profiles on various decorations such as baskets or fences. A tool was developed to identify potential

Content from this work may be used under the terms of the Creative Commons Attribution 3.0 licence. Any further distribution of this work must maintain attribution to the author(s) and the title of the work, journal citation and DOI.
Published under licence by IOP Publishing Ltd

redesigns for the assembly process. The tool is displayed according to its application in accordance with the actual data and compared with the decision of its interests (El-nounu, 2017).

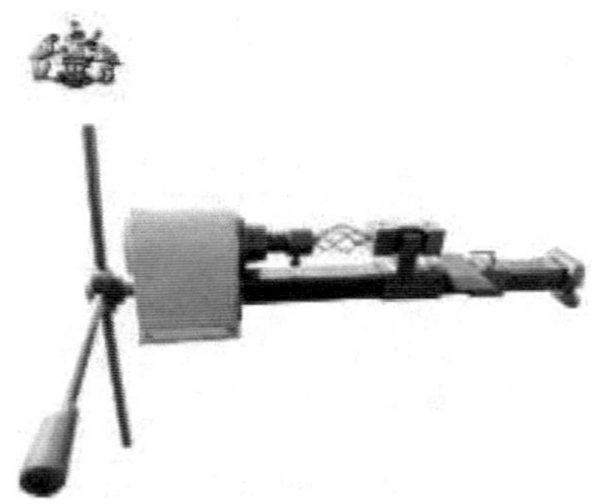

Figure 1. Existing tool

Figure 2. Component to twist

The product, is an artifact that can be designed and the design of the process is contained in a proposal that is descriptive and satisfies the needs (Rasoulifar et al., 2007). An important part of the sofa or chair that is produced is the frame of the chair itself. The shape of the chair frame for the product needed this time requires additional angles on the hollow pipe profile. To make the angle of the pipe needed twisting or twisting process of the pipe.

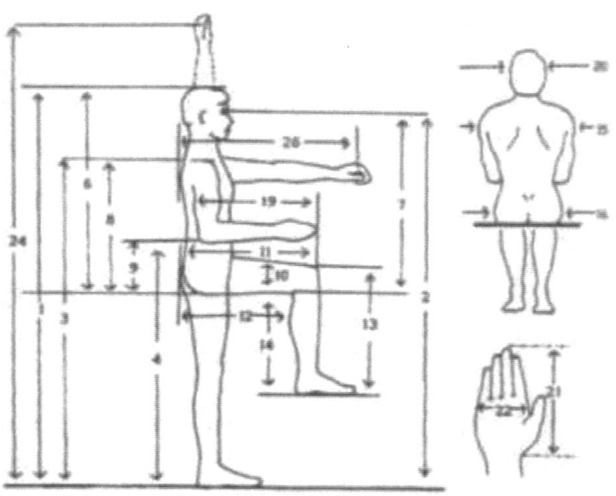

Figure 3. Anthropometri Indonesian People Size (Nurmanto, 2004)

According to Stevenson and Nurmianto in Febriani (2015), anthropometry is a numerical data set that relates to the characteristics of the human body in the form of size, shape and strength, and the application of these data for handling design problems. Anthropometry can be used to find the proportional position of pushing this tool.

IOP Conf. Series: Materials Science and Engineering 1175 (2021) 012019 doi:10.1088/1757-899X/1175/1/012019

2. Method

This study uses the Ulrich method in the process to get the best design as needed. Setiawan (2014) in his design research entitled Development of Bicycle Design for Post-Stroke Patients using the Ulrich method to select and determine the best concepts from the alternatives made, as well as research conducted by Sholeh (2018) in a study entitled Analysis and Design of Static Bicycles for Rehabilitation of Stroke Patients. In addition, Zahro (2019) in his research entitled Underwater Turbine Portable Design and Fabrication with the Savonius Turbine Model also uses the Ulrich method to compare 3 design concepts that help the writer determine the design according to need, which is then realized by fabrication. Research on the design of this twist tool will use this method for the process of selecting the best concept design. According to Ulrich and Eppinge (2005), the process of developing concepts according to the Ulrich method includes the following activities: Identification of needs, Concept preparation, Concept selection, Tool Making.

3. Details and Design Determination

3.1. Table of Tool Design Requirement

A list of tool design requirements is obtained by observing and interviewing. Observations are made directly by researchers when doing practical work or internships in the company. For interview respondents the specification of equipment requirements consisted of supervisors and production workers. From the interview, we get the data of tool needs and desires that are expected to be realized which can then be compiled a list of needs of this hollow pipe twist tool. Descriptions of these needs will be grouped into several criteria as in the following issues in conclution:

1. Fungtional
2. Strength and safe
3. Ergonomy or easy to use
4. Assembly or manufacturing
5. Maintenance

3.2. Fungtional

In this research process, because it uses the Ulrich method, prepare 3 alternative design concepts in which the three functions have been analyzed so that they work according to their needs.

3.2.1. Concept 1.

First concept has no limit on the twisting degrees of the tool because the twist process uses 8 small holders. It consists of three main parts of the tool, namely base, drum, and top. And for ease of taking components after a twist, this concept has one hole big enough. It will use trapezoidal thread for top clamping and toggle clamp for the base.

3.2.2. Concept 2.

Second concept has an unlimited twisting limit even though it is more than one turn. Consists of three main parts, namely base, drum and top. But for the top it does not consist of a plate like the concept 1, but from the connection of hollow pipes. To guide the twist process, this concept uses 4 medium holders. Same with the first concept, it will use trapezoidal thread for top clamping and toggle clamp *for* the base.

3.2.3. Concept 3.

Third concept has a tool twisting limit of 73 degrees. Consists of three main parts, namely base, drum and top. To guide the twist process, this concept uses a guide plate and for easy retrieval of components after a twist, this concept has holes on two sides. Clamping system both for top and base using toggle but in different type.

Figure 4. From left, 1st concept, 2nd concept, and 3rd concept

3.3. Clamping System

In the three concepts that were made, they basically had the same stress strength that is in accordance with the attached catalog. Taking into account the two types of clamp that are easy to find also the size that fits the tool, namely the toggle clamp type P100&P102 and Trapezoidal Thread M16. A toggle clamp is able to hold up to a maximum of 270 kg or equivalent to 2647.8 N, and for threads can withstand loading up to 91.1 kN or equivalent to 9289.61 kg. As for the torsion needed, after being simulated using the Fusion 360 software, the results require a minimum magnitude moment of 120 Nm.

3.4. Concept Determination

3.4.1. According to dimention.

The required size must be less than the size of the work table for this tool which is 2x1.5 meters long x width. The smaller the size of the design will be better because it is easily placed on a desk that has been available.

Table 1. Concept Dimensions

Concept	Dimention (mm)			Area (mm^2)
	Length	Wide	Height	
1	635	550	718	$349{,}25x10^3$
2	575	500	708	$287{,}5x10^3$
3	575	500	718	$287{,}5x10^3$

IOP Conf. Series: Materials Science and Engineering 1175 (2021) 012019 doi:10.1088/1757-899X/1175/1/012019

3.4.2. According to Operational.

Operational considering to it is the number of steps for normal use, security with the presence or absence of sharp edges on the instrument, and also anthropometry.

Table 2. Operational Analysis

	Concept 1	Concept 2	Concept 3
Steps of use	6	6	6
Risk sharp edge	2	6	-

Based on considerations of use, it also uses a simple anthropometric analysis according to the list of needs on the third point, namely ergonomics for the convenience of its use. For the height of the handle that is good according to calculations with anthropometric tables, it is found that at least the location of the handle is at 1338 mm, which is the average shoulder height of a man. Therefore, it can be seen in the table for the height of the tool closest to the proportional value is the concept with the lowest difference.

Table 3. Anthropometry

(in mm)	Tool height	Table height	Total height	Propotional height	Differenc e
Concept 1	680	600	1280	1338	58
Concept 2	677	600	1277	1338	61
Concept 3	682	600	1282	1338	56

3.4.3. According to Strength.

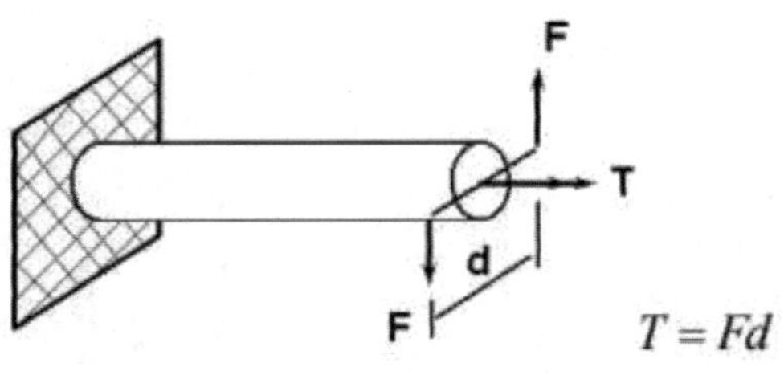

Consideration based on the force of thrust is done by calculating the torque with the principle as shown below. The strength needed to twist if it is assumed these concepts must twist by 10 degrees for aluminum components can be seen in the table.

Figure 4. Torque

$$T = F. \quad (1)$$

T = Torque (Nm)
F = Force (N)
D = Diameter (m)

IOP Conf. Series: Materials Science and Engineering 1175 (2021) 012019 doi:10.1088/1757-899X/1175/1/012019

Table 1. Strength torque analyse

[illegible]	[illegible]	[illegible]	[illegible]	[illegible]
[illegible]	[illegible]	[illegible]	[illegible]	[illegible]
[illegible]	[illegible]	[illegible]	[illegible]	[illegible]
[illegible]	[illegible]	[illegible]	[illegible]	[illegible]

3.4.4. According to Assembly.

Whereas in assembling, an analysis is carried out for the concept with the easiest and simplest type of work. And the fewer parts are considered to make it easier to make this tool.

Table 2. Assembly analyse

Concept	1	2	3
Number of Parts	85	76	63
Type of machining	2	3	2

3.4.5. According to Maintenance.

Considerations based on treatment were carried out and it resulted that the three concepts had an easy maintenance model of 3 points and differed in fairly difficult treatments with concept numbers 1,2, and 3 namely 6,7, and 5 points.

3.5.Best Design Determination

This study will do the concept selection with a decision matrix model with the stages of concept evaluation. Because there are only 3 alternative design concepts, filtering is not considered as necessary. For selection criteria that have been determined based on the specifications of the list of needs, there are dimensions, use, strength, assembly, and maintenance. To get results that are close to the selection criteria, analysis and 3D simulation is done on each alternative design concept model.

Table 6. Concept selection matrices

Selection Criterias	Precentage	Concept							
		Reference		1		2		3	
		Rate	Score	Rate	Score	Rate	Score	Rate	Score
Dimention	20%	4	0.8	4	0.8	4	0.8	4	0.8
Operational	15%	2	0.3	3	0.45	3	0.45	4	0.6
Strength	15%	2	0.3	4	0.6	4	0.6	4	0.6
Assembly	20%	2	0.4	3	0.6	2	0.4	4	0.8
Maintenance	15%	3	0.45	3	0.45	2	0.3	3	0.45
Absolut value	100%	13	2.25	17	2.9	15	2.55	17	3.25
Relative value			20.55		26.48		23.29		29.68
Ranking			4		2		3		1

Rate come from the considering before, and the score is the rate multiple by the precentage, the sum of the score would be the absolute value . Ranking got from the relative value number, and relative value number measured by the absolute value a concept of all the concepts. From the tabel we can conclude that the best design concept with the highest relative value percentage is the third concept.

4. Conclusion

This twist tool is made with a toggle clamp locking system at the bottom and also the top. The material used by this tool is Mild Steel ST42 and toggle clamp type P102 & P100. Then, for twisting by 10 degrees, the minimum load required to drive is 15 kg. For how to use the system, the first is to hold by locking the bottom of the tool by pushing the second toggle clamp. Then proceed with locking using a screw thread at the top. The twisting can then be done by pushing on the handle part clockwise or counterclockwise, as far as the angle that has been marked, with a note of adding an angle of 3 degrees. From the 3 concepts made, by analyzing according to each function then based on the assessment matrix that is by considering the dimensions, use, strength, assembly, and maintenance of tools, the best alternative concept chosen is third concept.

Related to the results of the experiment of the tool, it takes 3 additional degrees so for the angular need of an angle of 11 degrees the rotation needs to be done by 14 degrees due to the springback of aluminum material components. This needs to be studied and re-examined if further research is done. Similarly, the existence of powersource, also custom on the size of components that can be done by the tool can later be developed again for ease of use of the tool.

5. References

[1] Alexander C 1963 *Notes on the Synthesis of Form* (Cambridge, MA: Harvard University Press.Batan, I.M.L) 2012 Desain Produk, Ed Pertama (Guna Widya, Surabaya)
[2] Badan Pusat Statistika. *Pertumbuhan Produksi Industri Manufaktur Triwulan IV 2019.* www.bps.go.id/
[3] Batan I M L 2012 *Desain Produk* Ed Pertama (Guna Widya, Surabaya)
[4] El-Nounu, Rahma A, Popov A and Ratchev S 2017 Redesign methodology for mechanical assembly. *Research in Engineering Design* 10.1007/s00163-017-0255-6
[5] Giancoli D C 2001 *Fisika Jilid I* (terjemahan) (Jakarta: Penerbit Erlangga)
[6] National Aeronautics and Space Administration 1995 *MAN System Integration Standards* **1** 4 NASA-STD-3000 460B
[7] Nurmianto E 2008 *Ergonomi: Konsep Dasar dan Aplikasinya* Ed Kedua (Guna Widya, Surabaya, Indonesia)
[8] Prasidya, Y 2020 *Reform of Indonesia's Sluggish Manufacturing Sector Urgently Needed: Analysts* (Jakarta Post)
[9] Pressman R S 2002 *Rekayasa Perangkat Lunak* (ANDI, Yogyakarta)
[10] Rasoulifar dkk 2007 Proposal of a new design methodology in the surgical domain, DS 42: *Proceedings of ICED* 2007, *the 16th International Conference on Engineering Design* (Paris, France)
[11] Serway J W and Jewett Jr 2010 *Fisika Untuk Sains dan Teknik* (Jakarta: Salemba Teknika)
[12] Setiawan T A 2014 *Pengembangan Desain Sepeda untuk Pasien Pasca Stroke* (Surabaya: ITS)
[13] Sholeh A 2018 *Analisis dan Perancangan Sepeda Statis untuk Rehabilitasi Penderita Stroke* (Surabaya: PPNS)
[14] Stevenson M G 1989 *Principles of Ergonomics* (Center for Safety Science-University Of NSW, Australia)

[15] Ulrich K T and Eppinge S 2001 *Perancangan & Pengembangan Produk Salemba* (Jakarta: Salemba Teknika)
[16] Zahro S F 2019 *Desain dan Fabrikasi Underwater Turbine Portable dengan Model Turbin Savonius* (Surabaya: PPNS)

AUTHOR INDEX